Structures and Energies of Polycyclic Hydrocarbons

Springer
New York
Berlin
Heidelberg
Barcelona
Hong Kong
London
Milan
Paris
Singapore
Tokyo

Joan E. Shields

Structures and Energies
of Polycyclic Hydrocarbons

Springer

Joan E. Shields
Department of Chemistry
Long Island University
C.W. Post Campus
Brookville, NY 11548
USA
jes23@liu.edu

Library of Congress Cataloging-in-Publication Data
Shields, Joan E.
    Structures and energies of polycyclic hydrocarbons / Joan E. Shields.
        p.   cm.
    Includes bibliographical references and index.
    ISBN 0-387-95411-2 (softcover : acid-free paper)
        1. Polycyclic aromatic hydrocarbons.   2. Heat of formation.   I. Title.
    QD341.H9 S548 2001
    547′.61—dc21   2001057592

Printed on acid-free paper.

Production managed by MaryAnn Brickner; manufacturing supervised by Erica Bresler.
Photocomposed copy prepared from the author's electronic file.
Printed and bound by Sheridan Books, Inc., Ann Arbor, MI.
Printed in the United States of America.

9 8 7 6 5 4 3 2 1

ISBN 0-387-95411-2   SPIN 10861961

Springer-Verlag   New York Berlin Heidelberg
A member of BertelsmannSpringer Science+Business Media GmbH

To my nephew, Michael Alfano, for his patience and computer expertise and without whom this book would never have been completed.

CONTENTS

INTRODUCTION

This book is a compilation of calculated structural geometries and heats of formation of approximately 750 known saturated polycyclic hydrocarbons having from 5 to 15 carbon atoms and consisting of three to seven rings, i.e., C_5H_6 (tricyclopentanes) to $C_{15}H_{18}$ (heptacyclopentadecanes). Molecular mechanics provides a facile way to calculate not only the heat of formation of a molecule but also the geometry of the corresponding structure. The International Union of Pure and Applied Chemistry (IUPAC) and the *Chemical Abstracts* names of the compounds, along with any trivial names that have been published in the literature, have been supplied. A number of polycyclic hydrocarbons, whose syntheses have not yet been reported in the chemical literature, have been the subject of theoretical interest. Although many of these provide exciting synthetic challenges to organic chemists, only those compounds whose syntheses have been reported are included here. Also, the compounds presented are those polycyclic hydrocarbons that do not possess substituents or bonds that are not part of a ring.

Since the initial report of the synthesis of a given compound does not always provide the best synthetic method, and, in fact, later preparative schemes have often produced higher yields and more easily isolable products, literature references are not cited. Rather, the reader can obtain further information for each of the compounds and all references to any specific compound, including its synthesis, through the *Chemical Abstracts Service* registry numbers (CAS RNs). The methods used in the computation of structures, heats of formation, and nomenclature are discussed in subsequent sections. The compounds within each chapter are arranged by their IUPAC names.

MOLECULAR MECHANICS

Since Westheimer's[1] introduction of an empirical force field (molecular mechanics) in 1956, a number of different force fields have been developed and used successfully by organic chemists to predict the stability of compounds, to solve conformational and stereochemical problems, to study potential intermediates in organic reactions, and to predict vibrational spectra.

ix

Allinger's first molecular mechanics force field (MM1, 1971[2]) was subsequently improved and refined, resulting in MM2 in 1977,[3] MM3 in 1989,[4] and MM4 in 1996.[5] Allinger's method is presently the most widely used force field, especially for saturated hydrocarbons. The MM2 and MM3 programs are available commercially[6] and easily accessible on a desktop computer. This rapid empirical method for calculating thermodynamic properties and energy-minimized structures has been shown to give results as accurate as the best semi- and nonempirical quantum chemical calculations for saturated hydrocarbons. MM2 and MM3 were used in this book to compute the most stable conformational isomers and heats of formation of polycyclic hydrocarbons. The structures presented here are those calculated to be at their energy minimum.

The Molecular Mechanics method is based on a series of classical mechanical functions (force fields) of the nuclear positions in a molecule. These include a bond-stretching potential function (an approximation of the Morse potential energy curve), a bond-angle bending function, a torsional potential function that describes the change in potential energy resulting from varying a dihedral angle by rotating two bonds relative to each other around a third connected bond, and a nonbonded (van der Waals) potential function. The combination of these functions is the force field.

One assumption used in molecular mechanics calculations is that for each pair of atoms there are standard values for bond lengths, bond angles, and distances between nonbonded atoms and for which the corresponding potential functions are at a minimum. In addition, adjustable parameters are introduced into the potential functions expressions to provide the best fit of the calculated and experimental properties of compounds, such as heats of formation, conformational energies, and geometries. The potential functions and parameters are optimized by fitting experimental (vibrational, crystallographic, etc.) data obtained from simple compounds to give reasonable agreement between calculations and experiment. A second assumption used in molecular mechanics is that these parameters can be transferred to other compounds of similar structure. In this way, data from a small number of compounds are used to calculate the properties of a wide range of molecules. Since molecular mechanics involves only nuclear coordinates, it is particularly well suited to calculations of saturated compounds, where there are no resonance effects.

The sum of all potential functions in a force field is called the steric energy, i.e., the difference in energy between the actual molecule and a hypothetical molecule in which all structural features have their "ideal" values. Since the steric energy is a function of all potential functions (bond lengths, bond angles, torsional angles, and the distance between nonbonded atoms), it describes a conformational energy surface for the molecule. Steric energy does not represent a physical property of a molecule because its value depends on the force field used. Since different force fields use different potential functions and parameters, the steric energies calculated from two different force fields usually will not agree. However, the steric energy is used to calculate heats of formation that can be compared with experimental data. Also, within a given force field, steric energy provides a convenient way to compare the stability of conformational and geometric stereoisomers in which the same type of interaction is present. Thus, steric energy can be used directly for structure minimization and in conformational searches. An extensive discussion of Molecular Mechanics is beyond the scope of this book, but can be found in several excellent treatises on the subject.[7]

COMPUTATION OF MOLECULAR GEOMETRY

In this work the structures of the polycyclic hydrocarbons were drawn and their geometries minimized using PCModel,[8] a computational routine in which the user can construct structures that are subsequently subjected to a force field calculation. PCModel uses a modified MM2 force field, called MMX. For most hydrocarbons there is good agreement between MM2 and MMX. In the structure minimization process, bonds in the molecule are rotated and the coordinates of atoms are moved in the direction of steepest descent in the value of the steric energy until an energy minimum is obtained. However, the minimum energy molecular geometry resulting from structure minimization may not necessarily be the absolute (global) minimum conformation of the molecule, depending on the position of the original structure on the potential energy curve. For example, if the global minimum can only be obtained from the initial structure through a rotation that is energetically uphill, the minimization would proceed in the opposite direction and lead to the bottom of a local energy well. In this case, the minimization would result in a conformational structure corresponding to a saddle-point or local minimum.

Several algorithms are available for the global minimization of molecules. In this study GMMX[8] was used to perform a global search on the conformational surface of a molecule. GMMX (global-MMX) is a steric energy minimization program that uses the MMX force field to search conformational space and produces the lowest energy unique conformations. In this iterative computational method random bonds in the molecule are broken and the free segments are rotated prior to reattaching the bonds. If the bond breaking results in epimerization, the isomer is discarded. The process is repeated until no lower energy conformers are found. The lowest energy conformers from GMMX were then subjected to minimization by MM2 or MM3, and the resulting coordinate files were converted by Alchemy 2000[6] into the energy-minimized structures of the polycyclic hydrocarbons found in this book.

HEAT OF FORMATION

Heat of formation (ΔH_f^o) is a useful tool in structural chemistry as a predictor of thermodynamic stability. By definition, the heat of formation is the sum of the internal energy of a molecule (the energy of a hypothetical, motionless state with translational, rotational, and vibrational contributions) and a PV (pressure/volume) term. In molecular mechanics, heats of formation are calculated using a set of experimental or *ab-initio* (when experimental data is unavailable) bond enthalpy parameters. These are in the form of either bond increments or group increments. This calculation assumes that the heat of formation of the motionless state can be expressed as the sum of the heats of formation of the individual bonds or structural units, e.g., CH_3, CH_2, that constitute a molecule. Addition of the steric energy to this sum accounts for deformations in the geometry and leads to a larger heat of formation value. MM2 was used to calculate the heats of formation, except in those cases where the program did not provide the necessary parameters. In those cases, MM3 was employed. Compounds containing small rings, especially cyclopropane rings, are an example; the MM3 parameters for these compounds provide better agreement with experimental data.

NOMENCLATURE

The two most commonly used systems for naming organic compounds are trivial names and those derived from the IUPAC rules.[9] The *Chemical Abstracts* (CA) nomenclature system is rarely used in the chemical literature, probably due to the choice of names for ring systems and the difficulty in numbering the carbon atoms in polycyclic bridged systems. For example, a diamantane precursor (1a), whose IUPAC name is pentacyclo[8.3.1.0^{2,8}.0^{4,13}.0^{7,12}]tetradecane, is named dodecahydro 7,1,5-ethanylylideneacenaphthylene in the CA system, derived from the numbering shown in 1b.

1a   1b

Another unique characteristic of the CA system is the use of α and β to describe the stereochemistry of cyclic hydrocarbons,[10] and it probably is an additional impediment to using this system. In addition to the more common terminology to describe the stereochemistry of a compound, such as exo, endo, R, S, etc., for molecules having three or more stereogenic centers, α and β are used to describe ring positions in cyclic compounds. In unsubstituted polycyclic hydrocarbons, α is used to denote a hydrogen atom below the plane and β represents a hydrogen atom above the plane, usually designated in structures by dotted lines for an α hydrogen and

xiii

wedges for a β hydrogen atom. For example, the CA name of *anti*-pentacyclo[5.3.1.1^{2,6}.0^{1,7}.0^{2,6}]dodecane (2) is tetrahydro 1H,4H-dimethanocyclobuta[1,2:3,4]di-cyclopentene 3aα, 3bβ, 6aβ, 6bα. It is not surprising that chemists prefer to use the trivial name, *anti*-bis[3.2.1]propellane to describe this compound.

2

Even a cursory examination of the chemical literature will reveal confusion and errors in the names of organic compounds, especially the IUPAC names of bridged polycyclic hydrocarbons. The IUPAC rules for naming these hydrocarbons, based on the von Baeyer system, require that the main ring contain as many carbon atoms as possible, two of which will serve as bridge-head atoms for the main bridge, that the main bridge must be as large as possible, and that the superscripts locating the secondary bridges must be as small as possible. Since the choice of bridge-head carbon atoms is crucial to obtaining the correct IUPAC name, it should be expected that ring systems with several rings could result in incorrect names; this is evident in many instances in the chemical literature. The difficulty in using the IUPAC rules for naming polycyclic hydrocarbons and the subsequent proliferation of different numbering in the IUPAC names for the same compound probably led to the widespread use of trivial names in the organic chemical literature. Nickon and Silversmith's[11] book, *Organic Chemistry, The Name Game*, presents a fascinating and humorous study of the background and rationale for the use of trivial names to describe organic compounds. There is no dispute that it is more convenient to use the trivial names *churchane* or *homopentaprismane* to describe hexacyclo[5.4.0.0^{2,6}.0^{3,10}.0^{5,9}.0^{8,11}]un-decane (3) or *[20]fullerane* or *dodecahedrane*, rather than undecacyclo[9.9.0.0^{2,9}.0^{3,7}.0^{4,20}.0^{5,18}.0^{6,16}.0^{8,15}.0^{10,14}.0^{12,19}.0^{13,17}]eicosane for compound (4). Their trivial names provide a visualization of the structures of the compounds that is not obvious from the IUPAC name.

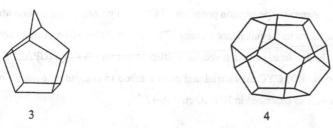

3   4

However, the use of less descriptive trivial names, such as nortwistane (5), garudane (6), 4-homobrendane (7), twistane (8), and christane (9), does present difficulties in visualizing structures.

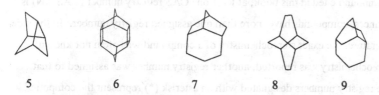

5   6   7   8   9

Compounding this problem is the use of different trivial names to describe the same compound, as well as various structural representations. For example, pentacyclo[6.4.0.0^{2,5}.0^{3,12}.0^{4,9}]dodecane (10) has been reported in the chemical literature under the trivial names ansarane, 1,3-bisethanocubane, [2.2.0]triblattane, and 3,10-dehydroditwistane, along with various structural illustrations, such as 10a and 10b.

10a   10b

In an effort to provide the reader with consistent and correct IUPAC names for the compounds cited in this book, and since many errors appear in the literature, these names were

xv

obtained through the use of two programs, TOPSYM and POLCYC, developed and described by Rucker and Rucker.[12] First, the carbon atoms are numbered arbitrarily and their connectivity is input to TOPSYM, a symmetry perception program. TOPSYM partitions the carbon atoms and their pairwise relationships into equivalence classes. The lowest numbered representative of each equivalence class of bridge-head pairs is saved for treatment in POLCYC, an IUPAC nomenclature program. POLCYC uses a trial and error method to compute the names of polycyclic hydrocarbons as described in IUPAC rule A-32.[9]

CHEMICAL ABSTRACTS REGISTRY NUMBERS

For every compound cited in this book, at least one CAS registry number (CAS RN) is given. In many instances, compounds have more than one assigned registry number. In the older organic chemical literature, the exact stereochemistry of a compound was often not known. Later, when the stereochemistry was reported, another registry number was assigned to that stereoisomer. Those registry numbers designated with an asterisk (*) represent the compound without any specified stereochemistry and may be accompanied by a registry number that represents the actual stereochemistry of the compound. Other designations include (±), (+), (-), (R), and (S), indicating the registry numbers for a racemic modification, the dextrorotatory, the levorotatory, and the R and S enantiomers, respectively.

REFERENCES

1. F. H. Westheimer, in *Steric Effects in Organic Chemistry*", M. S. Newman, Ed., Wiley, New York, 1956, Chapter 12.

2. N. L. Allinger, M. T. Tribble, M. A. Miller, and D. H. Wertz, *J. Am. Chem. Soc.,* **1971**, 93, 1637.

3. N. L. Allinger, *J. Am. Chem. Soc.,* **1977**, 99, 8127.

4. N. L. Allinger, Y. H. Yuh and J.-H. Lii, *J. Am. Chem. Soc.,* **1989**, 111, 8551.

5. N. L. Allinger, *J. Comp. Chem.,* **1996**. 17, 642.

6. MM2, MM3 and Alchemy 2000 are available from Tripos Associates, Inc., St. Louis, MO.

7. a) U. Burkett and N. L. Allinger, *Molecular Mechanics*, ACS Monographs 177, Washington, D. C., 1982; b) T. Clark, *A Handbook of Computational Chemistry: A Practical Guide to Chemical Structures and Energy Calculations*, Wiley-Interscience, New York, 1985; H. Dodzuik, *Modern Conformational Analysis*, VCH Publishers, Inc., New York, 1995.

8. Both PCModel and GMMX are available from Sabena Software, Bloomington, IN.

9. IUPAC Nomenclature of Organic Chemistry, 1957, *J. Am. Chem. Soc.,* **1960**, 82, 5545; 1979 Edition: International Union of Pure and Applied Chemistry, *Nomenclature of Organic Chemistry*, Sections A,B,C,D,E,F,H Pergamon Press. Oxford, 1979.

10. *Chemical Abstracts* Ring Index, Appendix IV, 1992, 194.

11. A. Nickon and E. F. Silversmith, *Organic Chemistry, The Name Game*, Pergamon Press, New York, 1987.

12. G. Rucker and C. Rucker, *Chimia,* **1990**, 44, 116; the programs were kindly supplied by Professor Rucker.

FORMAT

The polycyclic hydrocarbons in this book are arranged by the number of carbon atoms; each chapter contains all compounds with the same number of carbon atoms. The chapters are divided into sections corresponding to the molecular formula, e.g., Chapter 2, has three sections, namely, C_8H_{12}, tricyclooctanes; C_8H_{10}, tetracyclooctanes; and C_8H_8, pentacyclooctanes. The compounds are alphabetized by their IUPAC names, followed by the *Chemical Abstracts* registry number (CAS RN), the energy-minimized structure, any trivial names that have been used in the literature, the *Chemical Abstracts* name, and the heat of formation calculated by molecular mechanics (MM2 or MM3). An index of trivial names of the polycyclic hydrocarbons has been provided for the reader who wishes to learn more about the compound when only its trivial name is available.

An example of a pentacyclododecane shown below is illustrative of the format that is used in this book.

PENTACYCLO[6.4.0.0^{2,6}.0^{3,10}.0^{5,9}]DODECANE ← IUPAC NAME

CAS RN   150950-17-7   116347-39-8 (R) $\Big\rbrace$ CAS REGISTRY NUMBERS
              70209-47-1 (S)   70267-03-7 (±)

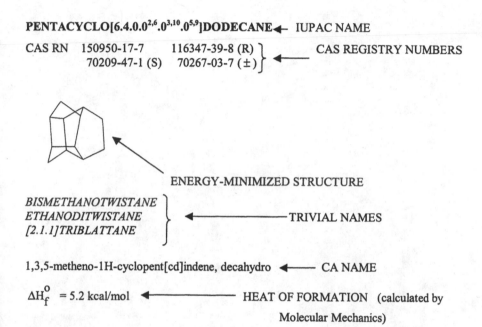

ENERGY-MINIMIZED STRUCTURE

BISMETHANOTWISTANE $\Big\rbrace$
ETHANODITWISTANE   ←——————— TRIVIAL NAMES
[2.1.1]TRIBLATTANE

1,3,5-metheno-1H-cyclopent[cd]indene, decahydro ←—— CA NAME

ΔH_f^o = 5.2 kcal/mol ←——————— HEAT OF FORMATION (calculated by

Molecular Mechanics)

1
POLYCYLCOPENTANES

A. TRICYCLOPENTANES - C_5H_6

TRICYCLO[1.1.1.0^{1,3}]PENTANE
CAS RN   35634-10-7

[1.1.1]PROPELLANE

$\Delta H_f^o = 103.5$ kcal/mol

TRICYCLO[1.1.1.0^{4,5}]PENTANE
CAS RN   333-17-5

$\Delta H_f^o = 119.3$ kcal/mol

TRICYCLO[2.1.0.0^{1,3}]PENTANE
CAS RN   78585-75-8

$\Delta H_f^o = 140.6$ kcal/mol

POLYCYCLOHEXANES
A. TRICYCLOHEXANES - C_6H_8

TRICYCLO[2.1.1.0^{1,4}]HEXANE
CAS RN   36120-91-9

[2.1.1]PROPELLANE

ΔH_f^o = 76.6 kcal/mol

TRICYCLO[2.2.0.0^{2,6}]HEXANE
CAS RN   327-63-9

ΔH_f^o = 74.0 kcal/mol

***cis-transoid-cis*-TRICYCLO[3.1.0.0^{2,4}]HEXANE**
CAS RN   21531-33-9
                311-83-1

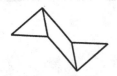

tricyclo[3.1.0.0^{2,4}]hexane 1α, 2β, 4β, 5α

ΔH_f^o = 40.4 kcal/mol

TRICYCLO[3.1.0.0^{2,6}]HEXANE
CAS RN 287-12-7

$\Delta H_f^O = 69.3$ kcal/mol

B. TETRACYCLOHEXANES - C$_6$H$_6$

TETRACYCLO[2.2.0.0^{2,6}.0^{3,5}]HEXANE
CAS RN   650-42-0

PRISMANE

ΔH_f^o = 86.9 kcal/mol

3
POLYCYCLOHEPTANES
A. TRICYCLOHEPTANES - C_7H_{10}

DISPIRO[2.0.2.1]HEPTANE
CAS RN   33475-22-8

ΔH_f^o = 74.9 kcal/mol

SPIRO[BICYCLO[2.1.0]PENTANE-2,1'-CYCLOPROPANE]
CAS RN   111850-98-7

ΔH_f^o = 103.2 kcal/mol

SPIRO[BICYCLO[2.1.0]PENTANE-5,1'-CYCLOPROPANE]
CAS RN   19446-68-5

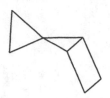

ΔH_f^o = 69.0 kcal/mol

TRICYCLO[2.2.1.0^{1,4}]HEPTANE
CAS RN   36120-90-8

[2.2.1]PROPELLANE

ΔH_f^o = 55.9 kcal/mol

TRICYCLO[2.2.1.0^{2,6}]HEPTANE
CAS RN   279-19-6

NORTRICYCLENE

ΔH_f^o = 19.8 kcal/mol

TRICYCLO[3.1.0.0^{2,4}]HEPTANE
CAS RN   36338-66-6

ΔH_f^o = 50.4 kcal/mol

TRICYCLO[3.1.1.0^{1,5}]HEPTANE
CAS RN   65513-21-5

[3.1.1]PROPELLANE

ΔH_f^o = 47.0 kcal/mol

TRICYCLO[3.1.1.0^{3,6}]HEPTANE
CAS RN   51273-50-8

NORCUBANE

ΔH_f^o = 52.9 kcal/mol

---

anti-**TRICYCLO[3.2.0.0^{2,4}]HEPTANE**
CAS RN   28102-61-1
                24592-19-6*

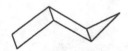

tricyclo[3.2.0.0^{2,4}]heptane 1α, 2β, 4β, 5α

ΔH_f^o = 56.9 kcal/mol

---

syn-TRICYCLO[3.2.0.0^{2,4}]HEPTANE
CAS RN   87304-84-5
             24592-19-6*

tricyclo[3.2.0.0^{2,4}]heptane 1α, 2α, 4α, 5α

ΔH_f^o = 88.1 kcal/mol

TRICYCLO[3.2.0.0^{2,7}]HEPTANE
CAS RN   279-18-5

ΔH_f^o = 38.3 kcal/mol

---

cis-TRICYCLO[4.1.0.0^{1,3}]HEPTANE
CAS RN   99416-71-4
             174-73-2*

cis-[3.5.3]FENESTRANE

ΔH_f^o = 45.7 kcal/mol

---

trans-TRICYCLO[4.1.0.0^{1,3}]HEPTANE

CAS RN   99416-70-3
              174-73-2*

trans-[3.5.3]FENESTRANE

ΔH_f^o = 54.6 kcal/mol

TRICYCLO[4.1.0.0^{1,5}]HEPTANE

CAS RN   52965-22-7

ΔH_f^o = 66.0 kcal/mol

---

cis-TRICYCLO[4.1.0.0^{2,4}]HEPTANE

CAS RN   16782-43-7
              187-26-8*

tricyclo[4.1.0.0^{2,4}]heptane 1α, 2α, 4α, 6α

ΔH_f^o = 29.4 kcal/mol

---

trans-TRICYCLO[4.1.0.0^{2,4}]HEPTANE
CAS RN   16782-44-8
              187-26-8*

tricyclo[4.1.0.0^{2,4}]heptane 1α, 2β, 4β, 6α

$\Delta H_f^o = 26.7$ kcal/mol

TRICYCLO[4.1.0.0^{2,7}]HEPTANE
CAS RN   287-13-8

$\Delta H_f^o = 46.8$ kcal/mol

B. TETRACYCLOHEPTANES - C_7H_8

TETRACYCLO[3.2.0.0^{1,6}.0^{2,6}]HEPTANE
CAS RN   109900-65-4

ΔH_f^o = 153.3 kcal/mol

TETRACYCLO[3.2.0.0^{2,7}.0^{4,6}]HEPTANE
CAS RN   278-06-8

QUADRICYCLANE

ΔH_f^o = 46.9 kcal/mol

TETRACYCLO[4.1.0.0^{2,4}.0^{3,5}]HEPTANE
CAS RN   50861-26-2

ΔH_f^o = 79.2 kcal/mol

4
POLYCYCLOOCTANES
A. TRICYCLOOCTANES - C$_8$H$_{12}$

DISPIRO[2.0.2.2]OCTANE
CAS RN   21426-37-9

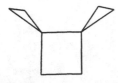

ΔH_f^o = 90.7 kcal/mol

DISPIRO[2.0.3.1]OCTANE
CAS RN   63783-86-8

ΔH_f^o = 84.0 kcal/mol

DISPIRO[2.1.2.1]OCTANE
CAS RN   25399-32-0

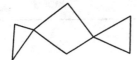

ΔH_f^o = 168.8 kcal/mol

TRICYCLO[2.2.2.0^{1,4}]OCTANE
CAS RN   36120-88-4

[2.2.2]PROPELLANE
ΔH_f^o = 46.1 kcal/mol

TRICYCLO[3.1.1.1^{2,4}]OCTANE
CAS RN   51273-49-5

DIASTERANE
ΔH_f^o = 52.0 kcal/mol

TRICYCLO[3.2.1.0^{1,5}]OCTANE
CAS RN   19074-25-0

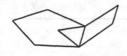

[3.2.1]PROPELLANE
ΔH_f^o = 32.2 kcal/mol

---

endo-TRICYCLO[3.2.1.0^{2,4}]OCTANE
CAS RN   278-72-8*

tricyclo[3.2.1.0^{2,4}]octane, 1α, 2α, 4α, 5α
ΔH$_f^o$ = 21.6 kcal/mol

exo-TRICYCLO[3.2.1.0^{2,4}]OCTANE
CAS RN   13377-46-3
            278-72-8*

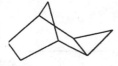

tricyclo[3.2.1.0^{2,4}]octane, 1α, 2β, 4β, 5α
ΔH$_f^o$ = 19.1 kcal/mol

TRICYCLO[3.2.1.0^{2,7}]OCTANE
CAS RN   285-43-8

ΔH$_f^o$ = 7.1 kcal/mol

TRICYCLO[3.2.1.0^{3,6}]OCTANE
CAS RN   250-22-6

ΔH_f^o = 10.0 kcal/mol

---

cis-cisoid-cis-**TRICYCLO[3.3.0.0^{2,4}]OCTANE**
CAS RN   22562-48-7
                24919-83-3*

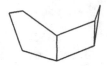

tricyclo[3.3.0.0^{2,4}]octane, 1α, 2α, 4α, 5α
ΔH_f^o = 34.4 kcal/mol

---

cis-transoid-cis-**TRICYCLO[3.3.0.0^{2,4}]OCTANE**
CAS RN   15774-40-0
                24919-83-3*

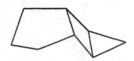

ΔH_f^o = 30.9 kcal/mol

TRICYCLO[3.3.0.0^{2,6}]OCTANE
CAS RN   250-21-5

ΔH_f^o = 7.1 kcal/mol

TRICYCLO[3.3.0.0^{2,7}]OCTANE
CAS RN   187-31-5

ΔH_f^o = 15.2 kcal/mol

TRICYCLO[3.3.0.0^{2,8}]OCTANE
CAS RN   2401-89-0

cyclopropa[cd]pentalene, octahydro
ΔH_f^o = 15.7 kcal/mol

TRICYCLO[3.3.0.0³,⁷]OCTANE
CAS RN   444-26-8

BISNORADAMANTANE
DINORADAMANTANE

ΔH_f^o = 14.8 kcal/mol

TRICYCLO[4.1.1.0¹,⁶]OCTANE
CAS RN   51273-56-4

[4.1.1]PROPELLANE

ΔH_f^o = 38.0 kcal/mol

TRICYCLO[4.1.1.0²,⁵]OCTANE
CAS RN   70970-60-4

ΔH_f^o = 37.9 kcal/mol

***trans*-TRICYCLO[4.2.0.0^{1,3}]OCTANE**
CAS RN 20589-03-1

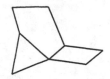

[3.5.4]FENESTRANE
tricyclo[4.2.0.0^{1,3}]octane, 3α, 6β

ΔH_f^o = 50.8 kcal/mol

TRICYCLO[4.2.0.0^{1,4}]OCTANE
CAS RN 75889-02-0

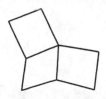

BROKEN WINDOW
[4.4.4]FENESTRANE

ΔH_f^o = 50.7 kcal/mol

TRICYCLO[4.2.0.0^{1,7}]OCTANE
CAS RN 53764-11-7

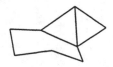

ΔH_f^o = 40.6 kcal/mol

trans-TRICYCLO[4.2.0.0^{2,4}]OCTANE
CAS RN   51088-76-8
              452-55-1*

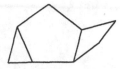

tricyclo[4.2.0.0^{2,4}]octane, 1α, 2β, 4β, 6α

ΔH_f^o = 30.0 kcal/mol

---

anti-TRICYCLO[4.2.0.0^{2,5}]OCTANE
CAS RN   13027-75-3
              277-08-7*

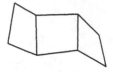

tricyclo[4.2.0.0^{2,5}]octane, 1α, 2β, 5β, 6α

ΔH_f^o = 46.4 kcal/mol

---

syn-TRICYCLO[4.2.0.0^{2,5}]OCTANE
CAS RN   28636-10-4
              277-08-7*

tricyclo[4.2.0.0^{2,5}]octane, 1α, 2α, 5α, 6α
all-syn-[3]LADDERANE

ΔH_f^o = 49.5 kcal/mol

TRICYCLO[4.2.0.0^{2,7}]OCTANE
CAS RN   98999-28-1

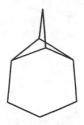

ΔH_f^o = 47.0 kcal/mol

TRICYCLO[4.2.0.0^{3,8}]OCTANE
CAS RN   3104-91-4

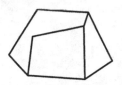

NORTWISTBRENDANE
9-NORTWISTBRENDANE

ΔH_f^o = 26.9 kcal/mol

***trans*-TRICYCLO[5.1.0.0^{1,3}]OCTANE**
CAS RN   10481-11-5

ΔH_f^o = 40.3 kcal/mol

---

cis-TRICYCLO[5.1.0.0^{2,4}]OCTANE

CAS RN   50695-42-6
            277–05–4*

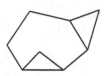

tricyclo[5.1.0.0^{2,4}]octane, 1α, 2α, 4α, 7α

ΔH_f^o = 25.5 kcal/mol

trans-TRICYCLO[5.1.0.0^{2,4}]OCTANE

CAS RN   50895-58-4
            277-05-4*

tricyclo[5.1.0.0^{2,4}]octane, 1α, 2β, 4β, 7α

ΔH_f^o = 42.8 kcal/mol

TRICYCLO[5.1.0.0^{2,8}]OCTANE

CAS RN   36328-29-7

ΔH_f^o = 60.2 kcal/mol

endo-TRICYCLO[5.1.0.0³,⁵]OCTANE
CAS RN   138284-15-8
              285-50-7*

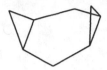

tricyclo[5.1.0.0³,⁵]octane, 1α, 3α, 5α, 7α

ΔH_f^o = 35.1 kcal/mol

---

exo-TRICYCLO[5.1.0.0³,⁵]OCTANE
CAS RN   62777-48-4
              285-50-7*

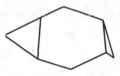

tricyclo[5.1.0.0³,⁵]octane, 1α, 3β, 5β, 7α

ΔH_f^o = 45.9 kcal/mol

B. TETRACYCLOOCTANES - C_8H_{10}

TETRACYCLO[3.2.0.0^{2,4}.0^{3,7}]OCTANE
CAS RN   4582-22-3

$\Delta H_f^o = 61.2$ kcal/mol

TETRACYCLO[3.2.1.0^{1,3}.0^{3,7}]OCTANE
CAS RN   81830-75-3

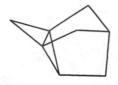

2,6-METHANO-2,6-DEHYDRONORBORNANE
$\Delta H_f^o = 56.3$ kcal/mol

cis-TETRACYCLO[3.3.0.0^{1,7}.0^{2,4}]OCTANE
CAS RN   140630-89-3
            141899-26-5*

tetracyclo[3.3.0.0^{1,7}.0^{2,4}]octane 2α, 4α, 5β, 7α
$\Delta H_f^o = 104.4$ kcal/mol

trans-TETRACYCLO[3.3.0.0^{1,7}.0^{2,4}]OCTANE
CAS RN   140438-59-1
            141899-26-5*

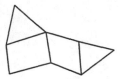

tetracyclo[3.3.0.0^{1,7}.0^{2,4}]octane, 2α, 4α, 5α, 7β

ΔH_f^o = 88.7 kcal/mol

TETRACYCLO[3.3.0.0^{2,4}.0^{3,6}]OCTANE
CAS RN   29185-91-9

ΔH_f^o = 58.0 kcal/mol

TETRACYCLO[3.3.0.0^{2,8}.0^{3,6}]OCTANE
CAS RN   5078-81-9

ΔH_f^o = 52.0 kcal/mol

TETRACYCLO[3.3.0.0^{2,8}.0^{4,6}]OCTANE
CAS RN   765-72-0

DIHYDROCUNEANE
SECOCUNEANE
HOMOQUADRICYLCENE
BISHOMOPRISMANE
dicyclopropa[cd,gh]pentalene, octahydro

ΔH_f^o = 28.0 kcal/mol

TETRACYCLO[4.2.0.0^{1,7}.0^{2,7}]OCTANE
CAS RN   98577-41-4

2,4-PROPANO[1.1.1]PROPELLANE

ΔH_f^o = 92.4 kcal/mol

TETRACYCLO[4.2.0.0^{2,4}.0^{3,5}]OCTANE
CAS RN   36328-44-6

ΔH_f^o = 90.5 kcal.mol

TETRACYCLO[4.2.0.0²,⁵.0³,⁸]OCTANE
CAS RN   3104-90-3
DIHYDROCUBANE

SECOCUBANE
ΔH_f^o = 84.4 kcal/mol

***exo*-TETRACYCLO[4.2.0.0²,⁸.0³,⁵]OCTANE**
CAS RN   78684-11-4
            57256-62-9*

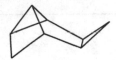

tetracyclo[4.2.0.0²,⁸.0³,⁵]octane 1α, 2β, 3β, 5β, 6β, 8β

ΔH_f^o = 112.0 kcal/mol

TETRACYCLO[5.1.0.0²,⁴.0³,⁵]OCTANE
CAS RN   55701-54-7

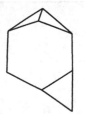

ΔH_f^o = 99.6 kcal/mol

C. PENTACYCLOOCTANES - C_8H_8

PENTACYCLO[3.3.0^{1,4}.0^{2,4}.0^{2,8}]OCTANE
CAS RN   122628-07-3

ΔH_f^o = 179.6 kcal/mol

PENTACYCLO[3.3.0.0^{2,4}.0^{3,7}.0^{6,8}]OCTANE
CAS RN   20656-23-9

CUNEANE

ΔH_f^o = 92.9 kcal/mol

PENTACYCLO[4.2.0.0^{2,5}.0^{3,8}.0^{4,7}]OCTANE
CAS RN   277-10-1

CUBANE
QUADRIPRISMANE
[4]PRISMANE

ΔH_f^o = 148.9 kcal/mol

PENTACYCLO[5.1.0.0^{2,4}.0^{3,5}.0^{6,8}]OCTANE
CAS RN   35434-67-4

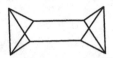

OCTABISVALENE

ΔH_f^o = 111.2 kcal/mol

5
POLYCYCLONONANES
A. TRICYCLONONANES - C_9H_{14}

DISPIRO[2.0.2.3]NONANE
CAS RN 24973-90-8

$\Delta H_f^o = 18.5$ kcal/mol

DISPIRO[2.1.2.2]NONANE
CAS RN 695-65-8

$\Delta H_f^o = 22.6$ kcal/mol

DISPIRO[3.0.3.1]NONANE
CAS RN 37677-07-9

$\Delta H_f^o = 66.4$ kcal/mol

SPIRO[BICYCLO[2.2.1]HEPTANE-2,1'-CYCLOPROPANE]
CAS RN   173-89-7

SPIRO[CYCLOPROPANE-1,2'-NORBORNANE]

ΔH_f^o = 8.2 kcal/mol

SPIRO[BICYCLO[2.2.1]HEPTANE-7,1'-CYCLOPROPANE]
CAS RN   159-41-1

CYCLOPROPYLIDENE NORBORNANE

ΔH_f^o = 12.9 kcal/mol

SPIRO[BICYCLO[3.2.0]HEPTANE-2,1'-CYCLOPROPANE]
CAS RN   38150-18-4

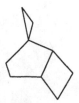

ΔH_f^o = 22.3 kcal/mol

SPIRO[BICYCLO[4.1.0]HEPTANE-3,1'-CYCLOPROPANE]
CAS RN   176790-96-8

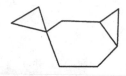

ΔH_f^o = 22.1 kcal/mol

TRICYCLO[3.2.1.1^{2,4}]NONANE
CAS RN   108471-09-6

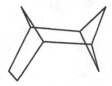

ΔH_f^o = 21.0 kcal/mol

TRICYCLO[3.2.2.0^{1,5}]NONANE
CAS RN   33107-49-2

[3.2.2]PROPELLANE
ΔH_f^o = 23.5 kcal/mol

TRICYCLO[3.2.2.0^{2,4}]NONANE
CAS RN   278-80-8

$\Delta H_f^o = 7.5$ kcal/mol

TRICYCLO[3.3.1.0^{1,5}]NONANE
CAS RN   27621-61-0

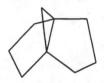

[3.3.1]PROPELLANE
1H,4H-3a, 6a-methanopentalene, tetrahydo

$\Delta H_f^o = 5.2$ kcal/mol

TRICYCLO[3.3.1.0^{2,4}]NONANE
CAS RN   28893-94-9

$\Delta H_f^o = 5.4$ kcal/mol

TRICYCLO[3.3.1.0²,⁷]NONANE
CAS RN   766-67-6

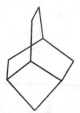

$\Delta H_f^o = 2.2$ kcal/mol

TRICYCLO[3.3.1.0²,⁸]NONANE
CAS RN   331-65-7

TETRAHYDROBARBARALANE

$\Delta H_f^o = 3.7$ kcal/mol

TRICYCLO[3.3.1.0³,⁷]NONANE
CAS RN   7075-86-7

NORADAMANTANE
2,5-methanopentalene, octahydro

$\Delta H_f^o = -14.2$ kcal/mol

TRICYCLO[4.2.1.0^{1,6}]NONANE
CAS RN   38325-64-3

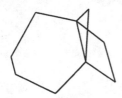

[4.2.1]PROPELLANE

$\Delta H_f^o = 20.5$ kcal/mol

***endo*-TRICYCLO[4.2.1.0^{2,4}]NONANE**
CAS RN   42733-02-8
                3928-79-8*

tricyclo[4.2.1.0^{2,4}]nonane 1α, 2α, 4α, 6α

$\Delta H_f^o = 8.5$ kcal/mol

***exo*-TRICYCLO[4.2.1.0^{2,4}]NONANE**
CAS RN   38655-74-2
                3928-79-8*

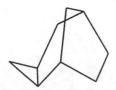

tricyclo[4.2.1.0^{2,4}]nonane 1α, 2β, 4β, 6α

$\Delta H_f^o = 6.1$ kcal/mol

---

endo-TRICYCLO[4.2.1.0^{2,5}]NONANE
CAS RN   16526-28-6
             249-81-0*

tricyclo[4.2.1.0^{2,5}]nonane 1α, 2α, 5α, 6α

ΔH_f^o = 13.2 kcal/mol

---

exo-TRICYCLO[4.2.1.0^{2,5}]NONANE
CAS RN   16526-27-5
             249-81-0*

tricyclo[4.2.1.0^{2,5}]nonane 1α, 2β, 5β, 6α

ΔH_f^o = 8.8 kcal/mol

TRICYCLO[4.2.1.0^{2,8}]NONANE
CAS RN   27197-55-3

ΔH_f^o = 7.4 kcal/mol

TRICYCLO[4.2.1.0³,⁷]NONANE
CAS RN   1521-75-1

DIHYDROHOMOCUNEANE
BRENDANE
1,5-methanopentalene, octahydro

$\Delta H_f^o = -10.5$ kcal/mol

TRICYCLO[4.3.0.0¹,³]NONANE
CAS RN   132974-79-9

cyclopropa[c]pentalene, octahydro

$\Delta H_f^o = 5.6$ kcal/mol

TRICYCLO[4.3.0.0²,⁴]NONANE
CAS RN   26472-28-6

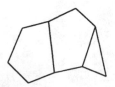

1H-cyclopropa[a]pentalene, octahydro

$\Delta H_f^o = 15.1$ kcal/mol

TRICYCLO[4.3.0.0^{2,7}]NONANE
CAS RN   108745-34-2

ΔH_f^o = 6.8 kcal/mol

TRICYCLO[4.3.0.0^{2,9}]NONANE
CAS RN   96308-35-9 (±)
                  3103-88-6*

1H-cycloprop[cd]indene, octahydro

ΔH_f^o = 10.0 kcal/mol

TRICYCLO[4.3.0.0^{3,7}]NONANE
CAS RN   60133-47-3 (-)
                  3104-87-8*

BREXANE
1,4-methanopentalene, octahydro 1α, 3aβ, 4α, 6aβ

ΔH_f^o = - 8.2 kcal/mol

TRICYCLO[4.3.0.0^{3,8}]NONANE
CAS RN   57287-49-7 (+)   42070-69-9(-)
             19026-97-2*   42070-81-5(±)

[1]TRIBLATTANE
NORTWISTANE
TWISTBRENDANE

ΔH_f^o = - 1.4 kcal/mol

***cis*-TRICYCLO[5.1.0.0^{2,6}]NONANE**
CAS RN   71055-12-4
             13913-22-9*

2,3-TRIMETHYLENEBICYCLO[2.1.1]HEXANE
2,3-(1,3-PROPANEDIYL)BICYCLO[2.1.1]HEXANE
cis-1,3-methanopentalene, octahydro

ΔH_f^o = 8.2 kcal/mol

TRICYCLO[5.2.0.0^{1,8}]NONANE
CAS RN   18220-66-1

ΔH_f^o = 37.9 kcal/mol

---

cis-TRICYCLO[6.1.0.0²,⁴]NONANE
CAS RN   81969-71-3
            277-83-8*

tricyclo[6.1.0.0²,⁴]nonane 1α, 2α, 4α, 8α

$\Delta H_f^o = 30.8$ kcal/mol

---

trans-TRICYCLO[6.1.0.0²,⁴]NONANE
CAS RN   81969-72-4
            277-83-8*

tricyclo[6.1.0.0²,⁴]nonane 1α, 2β, 4β, 8α

$\Delta H_f^o = 27.2$ kcal/mol

---

cis-TRICYCLO[6.1.0.0³,⁵]NONANE
CAS RN   285-53-0

$\Delta H_f^o = 27.6$ kcal/mol

B. TETRACYCLONONANES - C_9H_{12}

DISPIRO[CYCLOPROPANE-1,2'-BICYCLO[2.1.0]PENTANE-3',1''-CYCLOPROPANE]
CAS RN   72723-99-0

$\Delta H_f^o = 78.8$ kcal/mol

5,5'-SPIROBI[BICYCLO[2.1.0]PENTANE]
CAS RN   82482-48-2

$\Delta H_f^o = 101.1$ kcal/mol

SPIRO[CYCLOPENTANE-1,2'-TRICYCLO[1.1.1.0^{1,3}]PENTANE]
CAS RN   189825-08-9

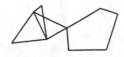

$\Delta H_f^o = 101.1$ kcal/mol

SPIRO[CYCLOPROPANE-1,3'-TRICYCLO[2.2.1.0^{2,6}]HEPTANE]
CAS RN   51370-31-1

NORTRICYCLENE-3-SPIROCYCLOPROPANE

ΔH_f^o = 38.8 kcal/mol

SPIRO[CYCLOPROPANE-1,3'-TRICYCLO[3.2.0.0^{2,7}]HEPTANE]
CAS RN   94348-12-6

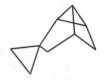

ΔH_f^o = 55.5 kcal/mol

SPIRO[CYCLOPROPANE-1,6'-TRICYCLO[3.2.0.0^{2,7}]HEPTANE]
CAS RN   24976-95-2

ΔH_f^o = 83.1 kcal/mol

SPIRO[CYCLOPROPANE-1,5'-TRICYCLO[4.1.0.0^{2,4}]HEPTANE]
CAS RN   76355-82-3

ΔH_f^o = 34.8 kcal/mol

TETRACYCLO[3.3.1.0^{2,4}.0^{2,8}]NONANE
CAS RN   53922-41-1

ΔH_f^o = 49.7 kcal/mol

TETRACYCLO[3.3.1.0^{2,4}.0^{3,7}]NONANE
CAS RN   20454-87-9

[3]PERISTYLANE
TRIAXANE
2,3-methanocyclopropa[cd]pentalene, octahydro

ΔH_f^o = 23.2 kcal/mol

---

endo, exo-**TETRACYCLO[3.3.1.0^{2,4}.0^{6,8}]NONANE**
CAS RN   55722-42-2
             187-49-5*

tetracyclo[3.3.1.0^{2,4}.0^{6,8}]nonane 1α, 2α, 4α, 5α, 6β, 8β

ΔH_f^o = 70.9 kcal/mol

---

exo, exo-**TETRACYCLO[3.3.1.0^{2,4}.0^{6,8}]NONANE**
CAS RN   24506-61-4
             187-49-5*

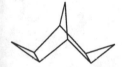

tetracyclo[3.3.1.0^{2,4}.0^{6,8}]nonane 1α, 2β, 4β, 5α, 6β, 8β

ΔH_f^o = 68.6 kcal/mol

TETRACYCLO[3.3.1.0^{2,8}.0^{4,6}]NONANE
CAS RN   3105-29-1

TRIASTERANE

ΔH_f^o = 28.0 kcal/mol

TETRACYCLO[4.2.1.0^{2,5}.0^{3,7}]NONANE
CAS RN   25557-71-5

1,3,5-methenopentalene, octahydro

$\Delta H_f^o = 32.2$ kcal/mol

TETRACYCLO[4.3.0.0^{1,5}.0^{2,6}]NONANE
CAS RN   124618-79-7

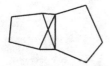

$\Delta H_f^o = 39.0$ kcal/mol

TETRACYCLO[4.3.0.0^{2,4}.0^{3,5}]NONANE
CAS RN   70938-88-4

1,2,3-methenopentalene, octahydro

$\Delta H_f^o = 64.7$ kcal/mol

TETRACYCLO[4.3.0.0^{2,4}.0^{3,7}]NONANE
CAS RN   6567-11-9

DELTACYCLANE
1,2,4-methenopentalene, octahydro

ΔH_f^o = 22.9 kcal/mol

TETRACYCLO[4.3.0.0^{2,4}.0^{3,8}]NONANE
CAS RN   60997-83-3

DEHYDROBRENDANE
DIHYDROHOMOCUNEANE
1,3-methanocyclopropa[cd]pentalene, octahydro

ΔH_f^o = 27.7 kcal/mol

TETRACYCLO[4.3.0.0^{2,5}.0^{3,8}]NONANE
CAS RN   100762-55-8
            19540-88-6*

DEHYDROTWISTBRENDANE
DIHYDROHOMOCUBANE
[1.0]TRIBLATTANE

ΔH_f^o = 40.6 kcal/mol

exo-TETRACYCLO[4.3.0.0^{2,9}.0^{3,5}]NONANE
CAS RN   55689-15-1

1H-dicyclopropa[a,cd]pentalene, octahydro 2aα, 2bα, 2cα, 3aα, 3bα, 3cβ

$\Delta H_f^o = 57.2$ kcal/mol

TETRACYCLO[4.3.0.0^{2,9}.0^{5,7}]NONANE
CAS RN   74515-98-3

1H-dicycloprop[cd,hi]indene, octahydro

$\Delta H_f^o = 29.1$ kcal/mol

TETRACYCLO[4.3.0.0^{3,9}.0^{4,7}]NONANE
CAS RN   56555-19-2

DIHYDROHOMOCUBANE

$\Delta H_f^o = 43.1$ kcal/mol

TETRACYCLO[6.1.0.0^{2,4}.0^{5,7}]NONANE
CAS RN   62447-30-7
                24609-59-4*

cis-tris-σ-HOMOBENZENE
TRIHOMOBENZENE
tetracyclo[6.1.0.0^{2,4}.0^{5,7}]nonane 1α, 2α, 4α, 5α, 7α, 8α

ΔH_f^o = 42.6 kcal/mol

TETRACYCLO[6.1.0.0^{2,4}.0^{5,7}]NONANE
CAS RN   37831-90-6
                24609-59-4*

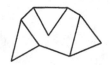

trans-tris-σ-HOMOBENZENE
TRIHOMOBENZENE
tetracyclo[6.1.0.0^{2,4}.0^{5,7}]nonane 1α, 2α, 4α, 5β, 7β, 8α

ΔH_f^o = 53.0 kcal/mol

TRISPIRO[2.0.0.2.1.1]NONANE
CAS RN   50874-24-3

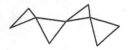

ΔH_f^o = 104.6 kcal/mol

TRISPIRO[2.0.2.0.2.0]NONANE
CAS RN   31561-59-8

TRICYCLOPROPYLIDENE
[3]ROTANE
[3.3]ROTANE

$\Delta H_f^o = 103.3$ kcal/mol

C. PENTACYCLONONANES - C_9H_{10}

PENTACYCLO[4.3.0.0^{1,7}.0^{2,9}.0^{7,9}]NONANE
CAS RN   122599-09-1

ΔH_f^o = 173.7 kcal/mol

PENTACYCLO[4.3.0.0^{2,4}.0^{3,8}.0^{5,7}]NONANE
CAS RN   13084-56-5

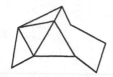

NORSNOUTANE
HOMOCUNEANE
1,2-methanodicyclopropa[cd,gh]pentalene, octahydro

ΔH_f^o = 90.7 kcal/mol

PENTACYCLO[4.3.0.0^{2,5}.0^{3,8}.0^{4,7}]NONANE
CAS RN   452-61-9

HOMOCUBANE
ΔH_f^o = 95.1 kcal/mol

PENTACYCLO[4.3.0.0^{2,8}.0^{3,5}.0^{4,7}]NONANE
CAS RN 60803-91-0

HOMOCUNEANE

ΔH_f^o = 87.3 kcal/mol

PENTACYCLO[5.2.0.0^{1,8}.0^{2,8}.0^{3,5}]NONANE
CAS RN 120546-07-8

ΔH_f^o = 165.7 kcal/mol

SPIRO[CYCLOPROPANE-1,3'-TETRACYCLO[3.2.0.0^{2,7}.0^{4,6}]HEPTANE]
CAS RN 25049-26-7

ΔH_f^o = 54.3 kcal/mol

6

POLYCYCLODECANES
A. TRICYCLODECANES - $C_{10}H_{16}$

DISPIRO[2.0.2.4]DECANE
CAS RN   24029-74-1

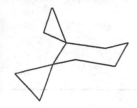

$\Delta H_f^o = 6.4$ kcal/mol

DISPIRO[2.2.2.2]DECANE
CAS RN   24518-94-3

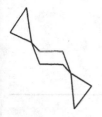

$\Delta H_f^o = 10.5$ kcal/mol

DISPIRO[3.1.3.1]DECANE
CAS RN   185-3-1

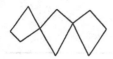

$\Delta H_f^o = 81.5$ kcal/mol

SPIRO[BICYCLO[2.2.1]HEPTANE-7,1'-CYCLOBUTANE]
CAS RN   53836-52-5

$\Delta H_f^o = 5.0$ kcal/mol

SPIRO[BICYCLO[2.2.2]OCTANE-2,1'-CYCLOPROPANE]
CAS RN   53764-10-6

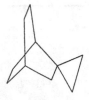

$\Delta H_f^o = -4.3$ kcal/mol

SPIRO[BICYCLO[3.1.0]HEXANE-6,1'-CYCLOPENTANE]
CAS RN   176-58-9

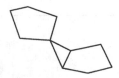

$\Delta H_f^o = 3.5$ kcal/mol

SPIRO[BICYCLO[3.2.1]OCTANE-6,1'-CYCLOPROPANE]
CAS RN   86359-34-4

ΔH_f^o = -3.0 kcal/mol

SPIRO[BICYCLO[3.3.0]OCTANE-2,1'-CYCLOPROPANE]
CAS RN   30492-96-7

spiro[cyclopropane-1,1'(2'H)-pentalene], hexahydro

ΔH_f^o = 5.2 kcal/mol

TRICYCLO[3.3.1.1^{3,7}]DECANE
CAS RN   281-23-2

ADAMANTANE

ΔH_f^o = -31.5 kcal/mol

TRICYCLO[3.3.2.0^{1,5}]DECANE
CAS RN   27613-46-3

[3.3.2]PROPELLANE
1H,4H-3a,6a-ethanopentalene, tetrahydro

ΔH_f^o = - 2.2 kcal/mol

TRICYCLO[3.3.2.0^{2,8}]DECANE
CAS RN   283-49-8

TRIHOMONORTRICYCLENE
HEXAHYDROBULLVALENE

ΔH_f^o = 6.8 kcal/mol

TRICYCLO[3.3.2.0^{3,7}]DECANE
CAS RN   49700-60-9

9-HOMONORADAMANTANE
1,9-HOMONORADAMANTANE
2,5-ethanopentalene, octahydro

ΔH_f^o = -17.3 kcal/mol

---

cis-cisoid-cis-TRICYCLO[4.2.1.1^{2,5}]DECANE
CAS RN   49700-69-8
             253-11-2*

syn-tricyclo[4.2.1.1^{2,5}]decane 1α, 2β, 5β, 6α

ΔH_f^o = - 0.7 kcal/mol

---

cis-transoid-cis-TRICYCLO[4.2.1.1^{2,5}]DECANE
CAS RN   49700-70-1
             253-11-2*

anti-tricyclo[4.2.1.1^{2,5}]decane 1α, 2α, 5α, 6α

ΔH_f^o = - 9.8 kcal/mol

TRICYCLO[4.2.2.0^{1,5}]DECANE
CAS RN   49700-63-2

1,7-TRIMETHYLENENORBORNANE
1,3a-ethanoperhydropentalene
1,3a-ethanopentalene, hexahydro

ΔH_f^o = - 10.1 kcal/mol

TRICYCLO[4.2.2.0^{1,6}]DECANE
CAS RN   31341-19-2

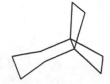

[4.2.2]PROPELLANE

ΔH_f^o = 16.5 kcal/mol

---

endo-**TRICYCLO[4.2.2.0^{2,5}]DECANE**
CAS RN   249-87-6*

ΔH_f^o = 26.6 kcal/mol

---

exo-**TRICYCLO[4.2.2.0^{2,5}]DECANE**
CAS RN   249-87-6*

TETRAHYDROBASKETENE

ΔH_f^o = 0.1 kcal/mol

TRICYCLO[4.3.1.0^{1,3}]DECANE
CAS RN   26845-38-5

ΔH_f^o = - 3.0 kcal/mol

TRICYCLO[4.3.1.0^{1,6}]DECANE
CAS RN   6049-83-8

[4.3.1]PROPELLANE
3a,7a-methano-1H-indene, hexahydro

ΔH_f^o = - 6.2 kcal/mol

TRICYCLO[4.3.1.0^{3,7}]DECANE
CAS RN   25107-18-0

ISOTWISTANE
1,5-methano-1H-indene, octahydro

ΔH_f^o = - 18.2 kcal/mol

TRICYCLO[4.3.1.0^{3,8}]DECANE
CAS RN   86022-59-5 (S)
              63902-00-1 (R)
              19026-94-9*

ISOADAMANTANE
PROTOADAMANTANE
2,5-methano-1H-indene, octahydro 2α, 3aβ, 5α, 7aβ

ΔH_f^o = − 19.7 kcal/mol

TRICYCLO[4.3.1.0^{8,10}]DECANE
CAS RN   53130-19-1

ΔH_f^o = 25.4 kcal/mol

---

cis-**TRICYCLO[4.4.0.0^{2,4}]DECANE**
CAS RN   73733-08-1

cycloprop[a]indene, decahydro

ΔH_f^o = - 2.4 kcal/mol

---

cis-TRICYCLO[4.4.0.0²·¹⁰]DECANE

CAS RN   95839-19-3
                53130-23-7*

cycloprop[de]naphthalene, decahydro 3aα, 3bα, 6aα, 6bα

ΔH_f^o = 3.7 kcal/mol

---

trans-TRICYCLO[4.4.0.0²·¹⁰]DECANE

CAS RN   95839-20-6
                53130-23-7*

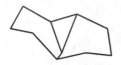

cycloprop[de]naphthalene, decahydro 3aα, 3bα, 6aβ, 6bα

ΔH_f^o = 18.6 kcal/mol

TRICYCLO[4.4.0.0³·⁸]DECANE

CAS RN   21449-14-9 (+)
                37165-27-8 (-)
                  253-14-5*

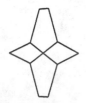

[2]TRIBLATTANE
TWISTANE

ΔH_f^o = - 16.2 kcal/mol

TRICYCLO[4.4.0.0^{3,9}]DECANE
CAS RN   29185-96-4

HOMOTWISTBRENDANE

ΔH_f^o = - 6.2 kcal/mol

***endo*-TRICYCLO[5.2.1.0^{1,5}]DECANE**
CAS RN   49700-58-5
             113724-88-2 (±)

endo-1,2-TRIMETHYLENENORBORNANE
3a,6-methano-3aH-indene, octahydro 3aα, 6α, 7aα

ΔH_f^o = - 6.4 kcal/mol

***exo*-TRICYCLO[5.2.1.0^{1,5}]DECANE**
CAS RN   49700-57-4
             113724-86-0 (±)

exo-1,2-TRIMETHYLENENORBORNANE
3a,6-methano-3aH-indene, octahydro 3aα, 6α, 7aβ

ΔH_f^o = - 16.0 kcal/mol

TRICYCLO[5.2.1.0^{1,6}]DECANE
CAS RN   172854-82-9

1,3a-methano-3aH-indene, octahydro 1α, 3aα, 7aβ

ΔH_f^o = 6.2 kcal/mol

TRICYCLO[5.2.1.0^{1,7}]DECANE
CAS RN   89857-39-6

[5.2.1]PROPELLANE

ΔH_f^o = 19.6 kcal/mol

endo-TRICYCLO[5.2.1.0^{2,4}]DECANE
CAS RN   51063-68-4
                 42578-04-1*

tricyclo[5.2.1.0^{2,4}]decane 1α, 2α, 4α, 7α

ΔH_f^o = 4.9 kcal/mol

exo-TRICYCLO[5.2.1.0^{2,4}]DECANE
CAS RN   51063-67-3
                42578-04-1*

tricyclo[5.2.1.0^{2,4}]decane 1α, 2β, 4β, 7α

ΔH_f^o = 5.4 kcal/mol

---

endo-TRICYCLO[5.2.1.0^{2,6}]DECANE
CAS RN   2825-83-4
                6004-38-2*

endo-2,3-TRIMETHYLENENORBORNANE
endo-TETRAHYDRODICYCLOPENTADIENE
4,7-methano-1H-indene, octahydro 3aα, 4α, 7α, 7aα

ΔH_f^o = - 13.4 kcal/mol

---

exo-TRICYCLO[5.2.1.0^{2,6}]DECANE
CAS RN   2825-82-3
                6004-38-2*

exo-2,3-TRIMETHYLENENORBORNANE
endo-TETRAHYDRODICYCLOPENTADIENE
4,7-methano-1H-indene, octahydro 3aα, 4β, 7β, 7aα

ΔH_f^o = - 16.3 kcal/mol

TRICYCLO[5.2.1.0³·⁸]DECANE
CAS RN   49700-65-4

4-HOMOBRENDANE
2-endo-6-endo-TRIMETHYLENENORBORNANE
2,4-methanoperhydro-1H-indene
2,4-methano-1H-indene, octahydro

ΔH_f^o = - 19.4 kcal/mol

TRICYCLO[5.2.1.0⁴·⁸]DECANE
CAS RN   42836-61-3

2-HOMOBRENDANE
1,6-methanoperhydroindene
1,6-methano-1H-indene, octahydro

ΔH_f^o = - 17.6 kcal/mol

all-cis-**TRICYCLO[5.2.1.0⁴·¹⁰]DECANE**
CAS RN   17760-91-7

TRIQUINACANE
PERHYDROQUINACENE
PERHYDROTRIQUINACENE
cyclopenta[cd]pentalene, decahydro

ΔH_f^o = - 22.1 kcal/mol

TRICYCLO[5.3.0.0^{1,5}]DECANE
CAS RN   101761-79-9

1H-cyclobuta[1,2:1,4]dicyclopentene, octahydro

ΔH_f^o = - 5.2 kcal/mol

***cis*-TRICYCLO[5.3.0.0^{2,4}]DECANE**
CAS RN   80339-56-5
             73733-07-0*

cycloprop[e]indene, decahydro 1aα, 3aβ, 6aβ, 6bα

ΔH_f^o = - 0.4 kcal/mol

***cis-cisoid-cis*-TRICYCLO[5.3.0.0^{2,6}]DECANE**
CAS RN   51607-14-8
             5650-12-4*

cyclobuta[1,2:3,4]dicyclopentene, decahydro 3aα, 3bα, 6aα, 6bα

ΔH_f^o = 1.3 kcal/mol

---

cis-transoid-cis-TRICYCLO[5.3.0.0^{2,6}]DECANE
CAS RN   36444-30-1
                      5650-12-4*

cyclobuta[1,2:3,4]dicyclopentene, decahydro 3aα, 3bβ, 6aβ, 6bα

ΔH_f^o = - 4.7 kcal/mol

TRICYCLO[5.3.0.0^{2,8}]DECANE
CAS RN   35834-42-5

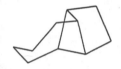

ΔH_f^o = 8.0 kcal/mol

TRICYCLO[5.3.0.0^{2,10}]DECANE
CAS RN   95839-21-7
                      3103-87-5*

cycloprop[cd]azulene, decahydro 2aα, 2bα, 6aα, 6bα

ΔH_f^o = 10.0 kcal/mol

TRICYCLO[5.3.0.0^{2,10}]DECANE
CAS RN   95839-22-8
                3103-87-5*

cycloprop[cd]azulene, decahydro 2aα, 2bα, 6aβ, 6bα

ΔH_f^o = 13.4 kcal/mol

TRICYCLO[5.3.0.0^{2,10}]DECANE
CAS RN   95839-23-9
                3103-87-5*

cycloprop[cd]azulene, decahydro 2aα, 2bβ, 6aα, 6bα

ΔH_f^o = 25.0 kcal/mol

TRICYCLO[5.3.0.0^{3,9}]DECANE
CAS RN   53130-27-1

4-HOMOTWISTBRENDANE
CRISTANE
2,5-TRIMETHYLENENORBORNANE

ΔH_f^o = - 6.2 kcal/mol

TRICYCLO[5.3.0.0⁴,⁸]DECANE

$$\text{TRICYCLO[5.3.0.0}^{4,8}\text{]DECANE}$$

CAS RN   19485-20-2

2-HOMOBREXANE

1,4-ethanopentalene, octahydro

ΔH_f^o = - 18.5 kcal/mol

---

cis-**TRICYCLO[6.1.1.0²,⁷]DECANE**

CAS RN   61899-07-8
           71016-21-2*

cis-1,3-methano-1H-indene, octahydro

ΔH_f^o = 5.5 kcal/mol

TRICYCLO[6.2.0.0¹,⁴]DECANE

CAS RN   71832-38-7

ΔH_f^o = 16.4 kcal/mol

---

cis-TRICYCLO[7.1.0.0^{1,3}]DECANE
CAS RN   151636-79-2
            152142-01-3*

ΔH_f^o = 34.4 kcal/mol

---

trans-TRICYCLO[7.1.0.0^{1,3}]DECANE
CAS RN   151587-92-7
            152142-01-3*

ΔH_f^o = 30.9 kcal/mol

---

trans-TRICYCLO[7.1.0.0^{2,4}]DECANE
CAS RN   81969-73-5
            31083-44-0*

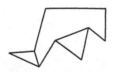

tricyclo[7.1.0.0^{2,4}]decane 1α, 2β, 4β, 9α
ΔH_f^o = 14.3 kcal/mol

---

cis-TRICYCLO[7.1.0.0^{3,5}]DECANE
CAS RN   589-58-2

ΔH_f^o = 20.2 kcal/mol

---

cis-TRICYCLO[7.1.0.0^{4,6}]DECANE
CAS RN   89955-58-8
286-73-7*

tricyclo[7.1.0.0^{4,6}]decane 1α, 4α, 6α, 9α

ΔH_f^o = 24.6 kcal/mol

---

trans-TRICYCLO[7.1.0.0^{4,6}]DECANE
CAS RN   89955-59-9
286-73-7*

tricyclo[7.1.0.0^{4,6}]decane 1α, 4β, 6β, 9α

ΔH_f^o = 23.5 kcal/mol

B. TETRACYCLODECANES - C₁₀H₁₄

DISPIRO[CYCLOBUTANE-1,2'-BICYCLO[1.1.0]BUTANE-4',1''-CYCLOBUTANE]
CAS RN   76308-11-7

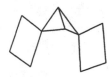

ΔH_f^o = 113.0 kcal/mol

DISPIRO[CYCLOPROPANE-1,2'-BICYCLO[2.2.0]HEXANE-3',1''-CYCLOPROPANE]
CAS RN   64141-42-0

ΔH_f^o = 77.2 kcal/mol

DISPIRO[CYCLOPROPANE-1,5'-BICYCLO[2.1.1]HEXANE-6',1''-CYCLOPROPANE]
CAS RN   89970-11-6

ΔH_f^o = 77.8 kcal/mol

SPIRO[CYCLOPROPANE-1,3'-TRICYCLO[3.2.1.0^{2,4}]OCTANE]
CAS RN   168037-71-6

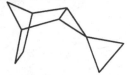

spiro[cyclopropane-1,3'-tricyclo[3.2.1.0^{2,4}]octane] 1'α, 2'β, 4'β, 5'α

ΔH_f^o = 52.5 kcal/mol

SPIRO[CYCLOPROPANE-1,6'-TRICYCLO[3.2.1.0^{2,4}]OCTANE]
CAS RN   86391-21-1

spiro[cyclopropane-1,6'-tricyclo[3.2.1.0^{2,4}]octane] 1'α, 2'β, 4'β, 5'α

ΔH_f^o = 47.2 kcal/mol

---

endo-SPIRO[CYCLOPROPANE-1,8'-TRICYCLO[3.2.1.0^{2,4}]OCTANE]
CAS RN   99396-94-8
          4117-71-9*

spiro[cyclopropane-1,8'-tricyclo[3.2.1.0^{2,4}]octane] 1'α, 2'β, 4'β, 5'α

ΔH_f^o = 55.0 kcal/mol

***exo*-SPIRO[CYCLOPROPANE-1,8'-TRICYCLO[3.2.1.0^{2,4}]OCTANE]**
CAS RN   99396-95-9

spiro[cyclopropane-1,8'-tricyclo[3.2.1.0^{2,4}]octane] 1'α, 2'α, 4'α, 5'α

ΔH_f^o = 57.8 kcal/mol

TETRACYCLO[3.3.1.1^{3,7}.0^{1,3}]DECANE
CAS RN   24569-89-9

1,3-DEHYDROADAMANTANE
1H,4H-2,5:3a,6a-dimethanopentalene, tetrahydro

ΔH_f^o = 17.1 kcal/mol

***endo, exo*-TETRACYCLO[3.3.2.0^{2,4}.0^{6,8}]DECANE**
CAS RN   63180-69-8
           27197-96-2*

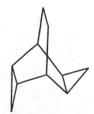

tetracyclo[3.3.2.0^{2,4}.0^{6,8}]decane 1α, 2α, 4α, 5α, 6β, 8β

ΔH_f^o = 54.8 kcal/mol

exo, exo-TETRACYCLO[3.3.2.0^{2,4}.0^{6,8}]DECANE
CAS RN   63180-71-2
                 27197-96-2*

tetracyclo[3.3.2.0^{2,4}.0^{6,8}]decane 1α, 2β, 4β, 5α, 6β, 8β

ΔH_f^o = 67.5 kcal/mol

TETRACYCLO[4.2.1.1^{2,5}.0^{1,6}]DECANE
CAS RN   52943-98-3
                 53797-19-6*

1,3-ETHANOBRIDGED[3.2.1]PROPELLANE
tetracyclo[4.2.1.1^{2,5}.0^{1,6}]decane 1α, 2α, 5α, 6α

ΔH_f^o = 42.2 kcal/mol

TETRACYCLO[4.3.1.0^{2,4}.0^{3,8}]DECANE
CAS RN   10501-16-3

2,4-DIDEHYDROADAMANTANE
2,4-DEHYDROADAMANTANE
2,4-methano-1H-cycloprop[cd]indene, octahydro

ΔH_f^o = 9.2 kcal/mol

TETRACYCLO[4.3.1.0^{2,9}.0^{3,5}]DECANE
CAS RN   71301-49-0

ΔH_f^o = 22.6 kcal/mol

TETRACYCLO[4.3.1.0^{2,9}.0^{4,8}]DECANE
CAS RN   31517-39-2

2,9-DEHYDRO-9-HOMONORADAMANTANE
5,7-DEHYDROPROTOADAMANTANE
1,4-methano-1H-cycloprop[cd]indene, octahydro

ΔH_f^o = 12.7 kcal/mol

TETRACYCLO[4.3.1.0^{3,8}.0^{7,9}]DECANE
CAS RN   71382-30-4

2,4-DEHYDRO-9-HOMONORADAMANTANE
SECOSNOUTANE
2,3-methano-1H-cycloprop[cd]indene, octahydro

ΔH_f^o = 16.4 kcal/mol

TETRACYCLO[4.4.0.0^{1,3}.0^{2,6}]DECANE
CAS RN 73373-39-4

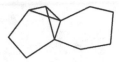

1, 3a, 7a-metheno-1H-indene, hexahydro

ΔH_f^o = 62.3 kcal/mol

TETRACYCLO[4.4.0.0^{1,7}.0^{2,7}]DECANE
CAS RN 108312-67-0

ΔH_f^o = 38.2 kcal/mol

TETRACYCLO[4.4.0.0^{2,4}.0^{3,5}]DECANE
CAS RN 70938-89-5

1,2,3-metheno-1H-indene, octahydro

ΔH_f^o = 60.8 kcal/mol

TETRACYCLO[4.4.0.0^{2,4}.0^{3,7}]DECANE
CAS RN   7203-36-3

2,10-DEHYDRO-4-HOMOBRENDANE
1,2,4-metheno-1H-indene, octahydro

ΔH_f^o = 12.6 kcal/mol

TETRACYCLO[4.4.0.0^{2,4}.0^{3,8}]DECANE
CAS RN   40002-42-4

DIHYDROSNOUTANE
1,3-methano-1H-cycloprop[cd]indene, octahydro

ΔH_f^o = 12.6 kcal/mol

TETRACYCLO[4.4.0.0^{2,5}.0^{3,8}]DECANE
CAS RN   100762-56-9
               266-89-7

[2.0]TRIBLATTANE
DIHYDROBASKETANE

ΔH_f^o = 27.8 kcal/mol

TETRACYCLO[4.4.0.0^{2,5}.0^{7,10}]DECANE
CAS RN   117858-06-7
                    4466-29-9*

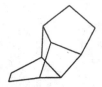

all-anti-[4]LADDERANE
PTERODACTYLANE
tetracyclo[4.4.0.0^{2,5}.0^{7,10}]decane 1α, 2β, 5β, 6α, 7β, 10β

ΔH_f^o = 111.6 kcal/mol

TETRACYCLO[4.4.0.0^{2,8}.0^{3,7}]DECANE
CAS RN   33234-23-0

1,4,7-methenoindane, hexahydro

ΔH_f^o = 16.9 kcal/mol

TETRACYCLO[4.4.0.0^{2,9}.0^{5,8}]DECANE
CAS RN   5603-26-9

DIHYDROBASKETANE
TETRAHYDROBASKETENE

ΔH_f^o = 47.0 kcal/mol

TETRACYCLO[4.4.0.0²,¹⁰.0⁴,⁸]DECANE
CAS RN   35856-12-3

5,10-DIDEHYDROPROTOADAMANTANE
5,10-DEHYDROPROTOADAMANTANE
2,5,6-metheno-1H-indene, octahydro

ΔH_f^o = 12.8 kcal/mol

TETRACYCLO[5.2.1.0¹,⁵.0⁵,⁹]DECANE
CAS RN   98640-30-3

4H-2,3b-methanocyclopropa[1,2:1,3]dicyclopentene, hexahydro

ΔH_f^o = 7.4 kcal/mol

---

endo, anti-**TETRACYCLO[5.2.1.0²,⁶.0³,⁵]DECANE**
CAS RN   53862-36-5
                252-55-1*

tetracyclo[5.2.1.0²,⁶.0³,⁵]decane 1α, 2α, 3β, 5β, 6α, 7α

ΔH_f^o = 38.1 kcal/mol

TETRACYCLO[5.2.1.0²,⁶.0³,⁸]DECANE
CAS RN   59054-55-6

ΔH_f^o = 12.5 kcal/mol

TETRACYCLO[5.2.1.0²,⁶.0⁴,⁸]DECANE
CAS RN   100762-57-0 (S)
            64727-80-6 (R)
            54445-73-7*

[8]DITWISTBRENDANE
BISNORDITWISTANE
BISNORTWISTANE
[1.1]TRIBLATTANE
8-DITWISTBRENDANE
1,5:2,4-dimethanopentalene, octahydro 1α, 2α, 3aβ, 4α, 5α, 6aβ

ΔH_f^o = 8.2 kcal/mol

TETRACYCLO[5.3.0.0¹,³.0⁴,⁶]DECANE
CAS RN   116316-74-6

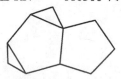

1H-dicyclopropa[ac]pentalene, octahydro

ΔH_f^o = 64.5 kcal/mol

TETRACYCLO[5.3.0.0^{2,6}.0^{3,9}]DECANE
CAS RN   85967-08-4

1,3-methano-1H-cyclobuta[cd]pentalene, octahydro

ΔH_f^o = 14.4 kcal/mol

TETRACYCLO[5.3.0.0^{2,6}.0^{3,10}]DECANE
CAS RN   34511-84-7

TETRAHYDROHYPOSTROPHENE

ΔH_f^o = 34.6 kcal/mol

TETRACYCLO[5.3.0.0^{2,10}.0^{4,8}]DECANE
CAS RN   144225-57-0

1,4-methano-1H-cyclopropa[a]pentalene, octahydro

ΔH_f^o = 10.5 kcal/mol

cis,cis-TETRACYCLO[7.1.0.0^{2,4}.0^{5,7}]DECANE
CAS RN   62279-40-7
               50510-95-7*

syn,syn-TRISHOMOCYCLOHEPTATRIENE
tetracyclo[7.1.0.0^{2,4}.0^{5,7}]decane 1α, 2α, 4α, 5α, 7α, 9α

ΔH_f^o = 59.3 kcal/mol

---

cis,trans-TETRACYCLO[7.1.0.0^{2,4}.0^{5,7}]DECANE
CAS RN   62279-36-1
               50510-95-7*

anti,syn-TRISHOMOCYCLOHEPTATRIENE
tetracyclo[7.1.0.0^{2,4}.0^{5,7}]decane 1α, 2α, 4α, 5β, 7β, 9α

ΔH_f^o = 56.3 kcal/mol

---

trans,trans-TETRACYCLO[7.1.0.0^{2,4}.0^{5,7}]DECANE
CAS RN   62279-35-0
               50510-95-7*

anti, anti-TRISHOMOCYCLOHEPTATRIENE
tetracyclo[7.1.0.0^{2,4}.0^{5,7}]decane 1α, 2β, 4β, 5α, 7α, 9α

ΔH_f^o = 50.5 kcal/mol

TRISPIRO[2.0.2.0.2.0.2.1]DECANE
CAS RN   25885-12-5

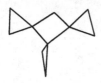

ΔH_f^o = 134.7 kcal/mol

C. PENTACYCLODECANES - $C_{10}H_{12}$

PENTACYCLO[4.4.0.0^{1,4}.0^{2,6}.0^{3,5}]DECANE
CAS RN   107575-49-5

ΔH_f^o = 94.9 kcal/mol

exo-PENTACYCLO[4.4.0.0^{2,4}.0^{3,7}.0^{8,10}]DECANE
CAS RN   126456-66-4 (R)
             16259-01-0*

2,3,5-metheno-1H-cyclopropa[a]pentalene, octahydro 1aα, 1bβ, 2α, 3α, 4aβ, 5α, 5aα

ΔH_f^o = 70.4 kcal/mol

PENTACYCLO[4.4.0.0^{2,4}.0^{3,8}.0^{5,7}]DECANE
CAS RN   28339-41-5

SNOUTANE
1,2,3-metheno-1H-cycloprop[cd]indene, octahydro

ΔH_f^o = 71.4 kcal/mol

PENTACYCLO[4.4.0.0^{2,4}.0^{3,9}.0^{5,7}]DECANE
CAS RN   35856-11-2

2,4,6,9-TETRADEHYDROADAMANTANE
1,2-methano-1H-dicycloprop[cd,hi]indene, octahydro

ΔH_f^o = 62.6 kcal/mol

PENTACYCLO[4.4.0.0^{2,5}.0^{3,8}.0^{4,7}]DECANE
CAS RN   5603-27-0

BASKETANE
1,1-BISHOMOCUBANE
1,8-BISHOMOCUBANE

ΔH_f^o = 83.8 kcal/mol

PENTACYCLO[4.4.0.0^{2,5}.0^{3,9}.0^{4,7}]DECANE
CAS RN   146450-22-8

1,2-BISHOMOCUBANE
1,2,5-metheno-1H-cyclobuta[cd]pentalene, octahydro

ΔH_f^o = 65.1 kcal/mol

PENTACYCLO[4.4.0.0^{2,5}.0^{3,9}.0^{4,8}]DECANE
CAS RN   7172-92-1

SECOPENTAPRISMANE
C$_{2V}$-BISHOMOCUBANE
1,3'-BISHOMOCUBANE

ΔH_f^o = 61.4 kcal/mol

PENTACYCLO[4.4.0.0^{2,8}.0^{3,5}.0^{4,7}]DECANE
CAS RN   60803-88-5

1,2,5-metheno-1H-cyclopropa[a]pentalene, octahydro

ΔH_f^o = 62.4 kcal/mol

PENTACYCLO[5.3.0.0^{2,4}.0^{3,6}.0^{5,8}]DECANE
CAS RN   77175-91-8

EVALANE I
1,4,5-metheno-1H-cyclopropa[a]pentalene, octahydro

ΔH_f^o = 68.6 kcal/mol

PENTACYCLO[5.3.0.0^{2,4}.0^{3,6}.0^{5,9}]DECANE
CAS RN   63858-72-0

DIHYDROBARETTANE
1,3,5-metheno-1H-cyclopropa[a]pentalene, octahydro

ΔH_f^o = 60.0 kcal/mol

PENTACYCLO[5.3.0.0^{2,5}.0^{3,9}.0^{4,8}]DECANE
CAS RN   62928-75-0 (-)
                 6707-86-4*

[1.1.0]TRIBLATTANE
C$_2$-BISHOMOCUBANE
1,3-HOMOCUBANE
1,3-BISHOMOCUBANE

ΔH_f^o = 47.8 kcal/mol

---

anti-**PENTACYCLO[5.3.0.0^{2,6}.0^{3,5}.0^{8,10}]DECANE**
CAS RN   61247-75-4
                 61665-03-0*

pentacyclo[5.3.0.0^{2,6}.0^{3,5}.0^{8,10}]decane 1α, 2β, 3α, 5α, 6β, 7α, 8β, 10β

ΔH_f^o = 107.2 kcal/mol

---

syn-**PENTACYCLO[5.3.0.0^{2,6}.0^{3,5}.0^{8,10}]DECANE**
CAS RN   61210-18-2
           61665-03-0*

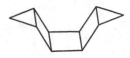

pentacyclo[5.3.0.0^{2,6}.0^{3,5}.0^{8,10}]decane 1α, 2α, 3β, 5β, 6α, 7α, 8β, 10β

ΔH_f^o = 108.2 kcal/mol

PENTACYCLO[5.3.0.0^{2,6}.0^{3,9}.0^{4,8}]DECANE
CAS RN   6707-88-6

1,4-BISHOMOCUBANE
1,3,4-metheno-1H-cyclobuta[cd]pentalene, octahydro

ΔH_f^o = 44.6 kcal/mol

PENTACYCLO[5.3.0.0^{2,10}.0^{3,5}.0^{6,8}]DECANE
CAS RN   27413-52-1

tricycloprop[cd,f,hi]indene, decahydro

ΔH_f^o = 37.7 kcal/mol

SPIRO[CYCLOPROPANE-1,1'-TETRACYCLO[3.3.0.0^{2,8}.0^{4,6}]OCTANE]
CAS RN   4019-68-5

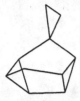

spiro[cyclopropane-1,1'(1'aH)-dicycloprop[cd,gh]pentalene], hexahydro

ΔH_f^o = 35.4 kcal/mol

D. HEXACYCLODECANES - $C_{10}H_{10}$

HEXACYCLO[4.4.0.0^{2,4}.0^{3,9}.0^{5,7}.0^{8,10}]DECANE
CAS RN   33840-23-2

DIADEMANE
MITRANE
CONGRESSANE
1,2,3-metheno-1H-dicycloprop[cd,hi]indene, octahydro

ΔH_f^o = 73.4 kcal/mol

HEXACYCLO[4.4.0.0^{2,4}.0^{3,9}.0^{5,8}.0^{7,10}]DECANE
CAS RN   41326-63-0

1,2,3-metheno-1H-cyclobuta[cd]cyclopropa[gh]pentalene, octahydro

ΔH_f^o = 105.5 kcal/mol

HEXACYCLO[4.4.0.0^{2,4}.0^{3,10}.0^{5,8}.0^{7,9}]DECANE
CAS RN   52674-19-8

BARETTANE
1,2,6:3,4,5-dimethenopentalene, octahydro

ΔH_f^o = 103.3 kcal/mol

HEXACYCLO[4.4.0.0^{2,5}.0^{3,9}.0^{4,8}.0^{7,10}]DECANE
CAS RN   4572-17-2

PENTAPRISMANE
[5]PRISMANE
HOUSANE

ΔH_f^o = 114.6 kcal/mol

7

POLYCYCLOUNDECANES
A. TRICYCLOUNDECANES $C_{11}H_{18}$

DISPIRO[2.0.2.5]UNDECANE
CAS RN   52879-54-6

ΔH_f^o = 6.5 kcal/mol

SPIRO[CYCLOPENTANE-1,2'-BICYCLO[2.2.1]HEPTANE]
CAS RN   172-61-2

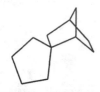

6,9-endo-METHYLENESPIRO[4.5]DECANE
SPIRO[CYCLOPENTANE-1,2'-NORBORNANE]

ΔH_f^o = - 20.2 kcal/mol

SPIRO[CYCLOPENTANE-1,2'-BICYCLO[4.1.0]HEPTANE]
CAS RN   180-42-7

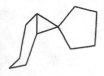

BICYCLO[4.1.0]HEPTANE-2-SPIROCYCLOPENTANE
SPIRO[CYCLOPENTANE-1,7'-NORCARANE]

ΔH_f^o = - 5.1 kcal/mol

TRICYCLO[3.3.3.0^{1,5}]UNDECANE
CAS RN   51027-89-5

[3.3.3]PROPELLANE
1H,4H-3a,6a-propanopentalene, tetrahydro

ΔH_f^o = - 28.3 kcal/mol

TRICYCLO[3.3.3.0^{2,4}]UNDECANE
CAS RN   42045-36-3

ΔH_f^o = 1.8 kcal/mol

TRICYCLO[4.2.2.1^{2,5}]UNDECANE
CAS RN   64822-53-3

ΔH_f^o = - 11.1 kcal/mol

endo-TRICYCLO[4.3.1.1^{2,5}]UNDECANE
CAS RN   64839-85-6
           55210-20-3*

2,4-endo-ETHANOBICYCLO[3.3.1]NONANE
tricyclo[4.3.1.1^{2,5}]undecane 1α, 2α, 5α, 6α

ΔH_f^o = - 11.8 kcal/mol

exo-TRICYCLO[4.3.1.1^{2,5}]UNDECANE
CAS RN   56245-92-2
           55210-20-3*

2,4-exo-ETHANOBICYCLO[3.3.1]NONANE
tricyclo[4.3.1.1^{2,5}]undecane 1α, 2β, 5β, 6α

ΔH_f^o = - 18.7 kcal/mol

TRICYCLO[4.3.1.1^{3,8}]UNDECANE
CAS RN   281-46-9

HOMOADAMANTANE

ΔH_f^o = - 27.2 kcal/mol

TRICYCLO[4.3.2.0^{1,5}]UNDECANE
CAS RN   64839-92-5

3a,7-ethano-3aH-indene, octahydro 3aα, 7α, 7aα

$\Delta H_f^o = -18.6$ kcal/mol

TRICYCLO[4.3.2.0^{1,5}]UNDECANE
CAS RN   64822-62-4

3a,7-ethano-3aH-indene, octahydro 3aα, 7α, 7aβ

$\Delta H_f^o = -24.4$ kcal/mol

TRICYCLO[4.3.2.0^{1,6}]UNDECANE
CAS RN   43043-80-7

[4.3.2]PROPELLANE
3a,7a-ethano-1H-indene, hexahydro

$\Delta H_f^o = -8.4$ kcal/mol

TRICYCLO[4.4.1.0^{1,6}]UNDECANE
CAS RN   6571-73-9

[4.4.1]PROPELLANE
4a,8a-methanonaphthalene, octahydro

ΔH_f^o = - 14.4 kcal/mol

TRICYCLO[4.4.1.0^{3,8}]UNDECANE
CAS RN   33657-52-2

2,6-methanonaphthalene, decahydro

ΔH_f^o = - 24.0 kcal/mol

TRICYCLO[5.2.2.0^{1,5}]UNDECANE
CAS RN   51095-23-9
           34894-52-5 (S)

3a,6-ethano-3aH-indene, octahydro

ΔH_f^o = - 25.1 kcal/mol

cis-TRICYCLO[5.2.2.0^{2,6}]UNDECANE
CAS RN   53432-45-4
                38255-97-9*

cis-2,3-TRIMETHYLENEBICYCLO[2.2.2]OCTANE
cis-4,7-ethano-1H-indene, octahydro

$\Delta H_f^o = -25.3$ kcal/mol

TRICYCLO[5.2.2.0^{4,8}]UNDECANE
CAS RN   7156-59-4

1,6-ethano-1H-indene, octahydro
1,6-ethanoindane, hexahydro

$\Delta H_f^o = -25.0$ kcal/mol

endo-TRICYCLO[5.3.1.0^{1,5}]UNDECANE
CAS RN   64839-91-4
                51027-88-4*

NORCEDRANE
1,7-*endo*-TRIMETHYLENEBICYCLO[3.2.1]OCTANE
1H-3a,7-methanoazulene, octahydro, 3aα, 7α, 8aα

$\Delta H_f^o = -16.6$ kcal/mol

exo-TRICYCLO[5.3.1.0^{1,5}]UNDECANE
CAS RN   55954-92-2
            51027-88-4*

NORCEDRANE
1,7-exo-TRIMETHYLENEBICYCLO[3.2.1]OCTANE
1H-3a,7-methanoazulene, octahydro, 3aα, 7α, 8aβ

ΔH_f^o = - 27.3 kcal/mol

---

cis, endo-TRICYCLO[5.3.1.0^{2,6}]UNDECANE
CAS RN   120052-75-7 (±)
            32743-61-6*

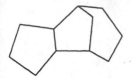

endo-6,7-TRIMETHYLENEBICYCLO[3.2.1]OCTANE
4,8-methanoazulene, decahydro 3aα, 4α, 8α, 8aβ

ΔH_f^o = - 6.9 kcal/mol

---

endo-TRICYCLO[5.3.1.0^{2,6}]UNDECANE
CAS RN   54676-38-9
            32743-61-6*

endo-6,7-TRIMETHYLENEBICYCLO[3.2.1]OCTANE
4,8-methanoazulene, decahydro 3aα, 4α, 8α, 8aα

ΔH_f^o = - 22.3 kcal/mol

---

exo-TRICYCLO[5.3.1.0²·⁶]UNDECANE
CAS RN   53495-28-6
                 32743-61-6*

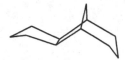

exo-6,7-TRIMETHYLENEBICYCLO[3.2.1]OCTANE
4,8-methanoazulene, decahydro 3aα, 4β, 8β, 8aα

ΔH_f^o = - 27.4 kcal/mol

TRICYCLO[5.3.1.0³·⁸]UNDECANE
CAS RN   43000-53-9

4-HOMOISOTWISTANE
2,6-TRIMETHYLENEBICYCLO[2.2.2]OCTANE
1,6-methanonaphthalene, decahydro

ΔH_f^o = - 28.5 kcal/mol

TRICYCLO[5.3.1.0³·⁹]UNDECANE
CAS RN   61770-04-5

4-HOMOPROTOADAMANTANE
2,5-methanoazulene, decahydro

ΔH_f^o = - 20.5 kcal/mol

---

cis-cisoid-cis-TRICYCLO[5.3.1.0^{3,11}]UNDECANE
CAS RN   83059-37-4 (±)

1H-cyclobuta[de]naphthalene, decahydro 1aα, 4aα, 7aα, 7bα

ΔH_f^o = - 7.3 kcal/mol

cis-transoid-cis-TRICYCLO[5.3.1.0^{3,11}]UNDECANE
CAS RN   83025-79-0

1H-cyclobuta[de]naphthalene, decahydro 1aα, 4aβ, 7aα, 7bα

ΔH_f^o = 3.2 kcal/mol

endo-TRICYCLO[5.3.1.0^{4,8}]UNDECANE
CAS RN   33566-66-4

endo-2,7-ETHANOBICYCLO[3.2.2]NONANE
1,6-methanoazulene, decahydro

ΔH_f^o = - 22.0 kcal/mol

TRICYCLO[5.3.1.0^{4,9}]UNDECANE
CAS RN   58008-62-1

2-HOMOPROTOADAMANTANE
2,7-methanonaphthalene, decahydro

$\Delta H_f^o = -23.9$ kcal/mol

TRICYCLO[5.3.1.0^{4,11}]UNDECANE
CAS RN   55820-81-0
           28099-09-4*

cis, endo-2,8-TRIMETHYLENEBICYCLO[3.3.0]OCTANE
PERHYDROCYCLOPENT[cd]INDENE
1H-cyclopent[cd]indene, decahydro 2aα, 4aα, 7aα, 7bα

$\Delta H_f^o = -26.9$ kcal/mol

TRICYCLO[5.3.1.0^{4,11}]UNDECANE
CAS RN   137330-58-6
              28099-09-4*

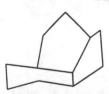

2,8-TRIMETHYLENEBICYCLO[3.3.0]OCTANE
PERHYDROCYCLOPENT[cd]INDENE
1H-cyclopent[cd]indene, decahydro 2aα, 4aα, 7aβ, 7bα

$\Delta H_f^o = -28.4$ kcal/mol

TRICYCLO[5.4.0.0^{1,3}]UNDECANE
CAS RN   99583-31-0

cyclopropa[d]naphthalene, decahydro 1aα, 4aα

ΔH_f^o = 8.0 kcal/mol

TRICYCLO[5.4.0.0^{1,3}]UNDECANE
CAS RN   99630-20-3

cyclopropa[d]naphthalene, decahydro 1aα, 4aβ

ΔH_f^o = - 12.6 kcal/mol

TRICYCLO[5.4.0.0^{1,5}]UNDECANE
CAS RN   6539-72-6

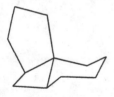

cyclopenta[1,4]cyclobuta[1,2]benzene, decahydro

ΔH_f^o = - 1.4 kcal/mol

TRICYCLO[5.4.0.0^{1,6}]UNDECANE
CAS RN   138736-71-7

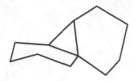

cyclopropa[1,2:1,3]dibenzene, decahydro

$\Delta H_f^o = -11.7$ kcal/mol

TRICYCLO[5.4.0.0^{2,9}]UNDECANE
CAS RN   4448-91-3

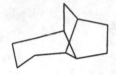

TETRAMETHYLENENORTRICYCLANE
1,4-methanoazulene, decahydro

$\Delta H_f^o = -17.7$ kcal/mol

trans-**TRICYCLO[5.4.0.0^{3,5}]UNDECANE**
CAS RN   20047-43-2
          94647-08-2*

2,3-METHANO-trans-DECALIN
1H-cyclopropa[b]naphthalene, decahydro 2aα, 6aβ

$\Delta H_f^o = 6.3$ kcal/mol

TRICYCLO[5.4.0.0³,⁸]UNDECANE
CAS RN   34650-87-8

1,5-methanonaphthalene, decahydro

ΔH_f^o = - 25.3 kcal/mol

TRICYCLO[5.4.0.0⁴,⁸]UNDECANE
CAS RN   60887-91-4
                   119972-04-2 (±)

2,4-BISHOMOBREXANE
2,9-ETHANOBICYCLO[3.3.1]NONANE
exo-2-syn-8-TRIMETHYLENEBICYCLO[3.2.1]OCTANE
1,4-ethano-1H-indene, octahydro

ΔH_f^o = - 25.1 kcal/mol

TRICYCLO[6.2.1.0¹,⁶]UNDECANE
CAS RN   55954-91-1
                   113724-89-3 (±)
                   74365-89-2*

2H-2,4a-methanonaphthalene, octahydro 2α, 4aα, 8aα

ΔH_f^o = - 24.8 kcal/mol

103

TRICYCLO[6.2.1.0^{1,6}]UNDECANE

CAS RN   36100-95-5
                113724-87-1 (±)
                74365-89-2*

2H-2,4a-methanonaphthalene, octahydro, 2α, 4aα, 8aβ

ΔH_f^o = - 26.6 kcal/mol

TRICYCLO[6.2.1.0^{2,5}]UNDECANE

CAS RN   444-08-6

ΔH_f^o = - 3.1 kcal/mol

---

cis, endo-TRICYCLO[6.2.1.0^{2,6}]UNDECANE

CAS RN   58116-65-7
                51027-86-1*

cis,endo-2,3-TRIMETHYLENEBICYCLO[3.2.1]OCTANE
4,7-methanoazulene, decahydro,3aα, 4β, 7β, 8aα

ΔH_f^o = - 24.4 kcal/mol

---

cis, exo-**TRICYCLO[6.2.1.0²,⁶]UNDECANE**
CAS RN   58116-66-8
                51027-86-2*

cis, exo-2,3-TRIMETHYLENEBICYCLO[3.2.1]OCTANE
4,7-methanoazulene, decahydro, 3aα, 4α, 7α, 8aα

ΔH_f^o = - 22.5 kcal/mol

---

trans, exo-**TRICYCLO[6.2.1.0²,⁶]UNDECANE**
CAS RN   64869-61-0
                51027-86-2*

trans, exo-2,3-TRIMETHYLENEBICYCLO[3.2.1]OCTANE
4,7-methanoazulene,decahydro 3aα, 4α, 7α, 8aβ

ΔH_f^o = - 22.8 kcal/mol

---

cis, endo-**TRICYCLO[6.2.1.0²,⁷]UNDECANE**
CAS RN   54676-30-1
                28691-42-1*

endo-2,3-TETRAMETHYLENEBICYCLO[2.2.1]HEPTANE
endo-2,3-TETRAMETHYLENENORBORNANE
1,4-methanonaphthalene, decahydro 1α, 4α, 4aα, 8aα

ΔH_f^o = - 19.8 kcal/mol

---

cis, exo-**TRICYCLO[6.2.1.0^{2,7}]UNDECANE**
CAS RN   32789-29-0
                28691-42-1*

exo-2,3-TETRAMETHYLENEBICYCLO[2.2.1]HEPTANE
exo-2,3-TETRAMETHYLENENORBORNANE
1,4-methanonaphthalene, decahydro 1α, 4α, 4aβ, 8aβ

ΔH_f^o = - 21.9 kcal/mol

TRICYCLO[6.2.1.0^{4,9}]UNDECANE
CAS RN   51027-87-3

2,4-BISHOMOBRENDANE
1,7-methanonaphthalene, decahydro

ΔH_f^o = - 25.7 kcal/mol

TRICYCLO[6.3.0.0^{1,4}]UNDECANE
CAS RN   119971-87-8 (±)

cyclobut[d]indene, decahydro 2aα, 5aβ

ΔH_f^o = - 4.7 kcal/mol

---

cis, cis-TRICYCLO[6.3.0.0^{1,5}]UNDECANE

CAS RN 61950-20-7 120052-65-5
55925-58-1*

1,2-exo-TRIMETHYLENE-cis-BICYCLO[3.3.0]OCTANE
cyclopenta[c]pentalene, decahydro

ΔH_f^o = - 26.8 kcal/mol

---

cis-cisoid-cis-TRICYCLO[6.3.0.0^{2,6}]UNDECANE

CAS RN 58116-67-9
6053-75-4*

1,3-TERNIPENTALANE
1H-cyclopenta[a]pentalene, decahydro 3aα, 3bα, 6aα, 7aα

ΔH_f^o = - 23.8 kcal/mol

---

cis-transoid-cis-TRICYCLO[6.3.0.0^{2,6}]UNDECANE

CAS RN 64839-76-5 120052-74-6
6053-75-4*

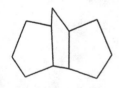

1,3-TERNIPENTALANE
1H-cyclopenta[a]pentalene, decahydro 3aα, 3bβ, 6aβ, 7aα

ΔH_f^o = - 25.0 kcal/mol

TRICYCLO[7.2.0.0^{2,5}]UNDECANE
CAS RN   126348-67-2

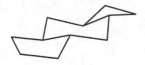

1H-cyclobut[f]indene, decahydro

ΔH_f^o = - 2.8 kcal/mol

---

cis-transoid-cis-**TRICYCLO[8.1.0.0^{3,5}]UNDECANE**
CAS RN   70518-96-6

tricyclo[8.1.0.0^{3,5}]undecane 1α, 3β, 5β, 10α

ΔH_f^o = 20.8 kcal/mol

B. TETRACYCLOUNDECANES - C₁₁H₁₆

DISPIRO[CYCLOPROPANE-1,2'-BICYCLO[2.2.1]HEPTANE-7,1''-CYCLOPROPANE]
CAS RN   65355-78-4

ΔH_f^o = 37.0 kcal/mol

DISPIRO[CYCLOPROPANE-1,2'-BICYCLO[2.2.1]HEPTANE-3',1''-CYCLOPROPANE]
CAS RN   40827-29-0

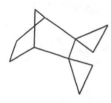

ΔH_f^o = 25.2 kcal/mol

***endo*-SPIRO[CYCLOPROPANE-1,6'-TRICYCLO[3.2.2.0²,⁴]NONANE]**
CAS RN   107079-24-3

spiro[cyclopropane-1,6'-tricyclo[3.2.2.0²,⁴]nonane 1'α, 2'β, 4'β, 5'α

ΔH_f^o = 32.9 kcal/mol

---

exo-SPIRO[CYCLOPROPANE-1,6'-TRICYCLO[3.2.2.0^{2,4}]NONANE]
CAS RN   107079-23-2

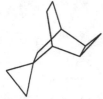

spiro[cyclopropane-1,6'-tricyclo[3.2.2.0^{2,4}]nonane 1'α, 2'α, 4'α, 5'α

ΔH_f^o = 32.9 kcal/mol

TETRACYCLO[4.2.2.1^{2,5}.0^{1,6}]UNDECANE
CAS RN   31341-18-1

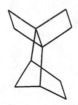

ΔH_f^o = 35.3 kcal/mol

TETRACYCLO[4.3.1.1^{4,8}.0^{1,3}]UNDECANE
CAS RN   88084-84-8

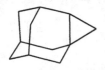

1,2-METHANOADAMANTANE

ΔH_f^o = 17.2 kcal/mol

TETRACYCLO[4.3.1.1^{4,8}.0^{1,4}]UNDECANE
CAS RN   26525-13-3

3,6-DEHYDROHOMOADAMANTANE
1H,4H-3a,6a-ethano-2,5-methanopentalene, tetrahydro

ΔH_f^o = 9.8 kcal/mol

TETRACYCLO[4.4.1.0^{2,4}.0^{3,9}]UNDECANE
CAS RN   170130-60-6

2,9-DEHYDROHOMOADAMANTANE
9,10-DEHYDROHOMOADAMANTANE
2,4-methanocycloprop[cd]azulene, decahydro

ΔH_f^o = 8.6 kcal/mol

TETRACYCLO[4.4.1.0^{3,11}.0^{9,11}]UNDECANE
CAS RN   144900-84-5
             144830-52-4
             106051-70-1*

[4.4.5.5]FENESTRANE
1,7-methano-1H-cyclobuta[c]pentalene, octahydro 1α, 2aα, 4aβ, 6aα

ΔH_f^o = 26.4 kcal/mol

111

TETRACYCLO[5.3.1.0^{1,7}.0^{4,11}]UNDECANE
CAS RN   54008-03-6

5H-cyclopenta[1,3]cyclopropa[1,2,3-cd]pentalene, octahydro

ΔH_f^o = 6.6 kcal/mol

TETRACYCLO[5.3.1.0^{2,4}.0^{3,9}]UNDECANE
CAS RN   71129-58-3

2,11-DEHYDROHOMOADAMANTANE
2,5-methanocycloprop[cd]azulene, decahydro

ΔH_f^o = 19.9 kcal/mol

---

endo-TETRACYCLO[5.3.1.0^{2,4}.0^{5,9}]UNDECANE
CAS RN   52746-12-0

4,5-endo-METHYLENEPROTOADAMANTANE
2,5-methanocycloprop[e]indene, decahydro 1aα, 2β, 3aα, 5β, 6aα, 6bα

ΔH_f^o = 14.6 kcal/mol

---

exo-TETRACYCLO[5.3.1.0^{2,4}.0^{5,9}]UNDECANE
CAS RN   52719-65-0

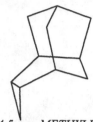

4,5-exo-METHYLENEPROTOADAMANTANE
2,5-methanocycloprop[e]indene, decahydro 1aα, 2α, 3aβ, 5α, 6aβ, 6bα

ΔH_f^o = 11.8 kcal/mol

TETRACYCLO[5.3.1.0^{2,5}.0^{3,9}]UNDECANE
CAS RN   55638-02-3

2,5-DEHYDROHOMOADAMANTANE
2,3-METHANO-2,4-DIDEHYDROADAMANTANE
2,5-methanocyclobut[cd]indene, decahydro

ΔH_f^o = 5.5 kcal/mol

TETRACYCLO[5.3.1.0^{2,5}.0^{4,9}]UNDECANE
CAS RN   59014-95-8

2,4-METHANOADMANTANE

ΔH_f^o = 3.7 kcal/mol

TETRACYCLO[5.3.1.0^{2,6}.0^{3,9}]UNDECANE
CAS RN   59014-96-9

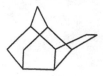

2,4-ETHANONORADAMANTANE
1,4-methanocyclopenta[cd]pentalene, decahydro

ΔH_f^o = -12.3 kcal/mol

TETRACYCLO[5.3.1.0^{2,6}.0^{4,9}]UNDECANE
CAS RN   58008-54-1

NORICEANE
1,6:2,5-dimethano-1H-indene, octahydro

ΔH_f^o = - 8.2 kcal/mol

***endo, exo*-TETRACYCLO[5.3.1.0^{2,6}.0^{8,10}]UNDECANE**
CAS RN   113932-76-6

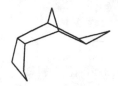

2,6-methanocycloprop[f]indene, decahydro 1aα, 2β, 3aβ, 5aβ, 6β, 6aα

ΔH_f^o = 16.4 kcal/mol

***exo, exo*-TETRACYCLO[5.3.1.0^{2,6}.0^{8,10}]UNDECANE**
CAS RN   114028-42-1

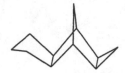

2,6-methanocycloprop[f]indene, decahydro 1aα, 2β, 3aα, 5aα, 6β, 6aα

ΔH_f^o = 17.3 kcal/mol

TETRACYCLO[5.3.1.0^{3,5}.0^{4,9}]UNDECANE
CAS RN   28786-66-5

2,4-DEHYDROHOMOADAMANTANE
2,5-methanocyclopropa[de]naphthalene, decahydro

ΔH_f^o = 1.2 kcal/mol

TETRACYCLO[5.3.1.0^{3,5}.0^{4,11}]UNDECANE
CAS RN   54008-04-7

1H-benzo[cd]cyclopropa[gh]pentalene, decahydro

ΔH_f^o = 9.4 kcal/mol

---

endo-TETRACYCLO[5.3.1.0^{4,11}.0^{8,10}]UNDECANE
CAS RN   67337-90-0

1H-cyclopenta[cd]cyclopropa[a]pentalene, octahydro 2aα, 4aα, 4bβ, 5aβ, 5bα, 5cα

ΔH_f^o = 26.7 kcal/mol

---

exo-TETRACYCLO[5.3.1.0^{4,11}.0^{8,10}]UNDECANE
CAS RN   55756-69-9

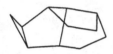

1H-cyclopenta[cd]cyclopropa[a]pentalene, octahydro 2aα, 4aα, 4bα, 5aα, 5bα, 5cα

ΔH_f^o = 24.2 kcal/mol

TETRACYCLO[5.4.0.0^{1,3}.0^{2,7}]UNDECANE
CAS RN   20345-00-0

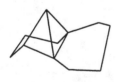

1,4a,8a-methenonaphthalene, octahydro

ΔH_f^o = 31.7 kcal/mol

TETRACYCLO[5.4.0.0^{1,6}.0^{2,7}]UNDECANE
CAS RN 98721-40-5

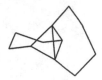

ΔH_f^o = 30.0 kcal/mol

TETRACYCLO[5.4.0.0^{2,4}.0^{3,9}]UNDECANE
CAS RN 42916-94-9

2,4-DEHYDRO-4-HOMOTWISTANE
1,4-methanocyclopropa[de]naphthalene, decahydro

ΔH_f^o = 8.6 kcal/mol

TETRACYCLO[5.4.0.0^{2,10}.0^{3,9}]UNDECANE
CAS RN 100693-72-9
100762-59-2 (±)

[3.0]TRIBLATTANE
ΔH_f^o = 27.0 kcal/mol

TETRACYCLO[5.4.0.0³,¹⁰.0⁴,⁸]UNDECANE
CAS RN 59014-98-1a

2,9-ETHANONORADAMANTANE
4,1,6-ethanylylidene-1H-indene, octahydro

ΔH_f^o = - 5.3 kcal/mol

TETRACYCLO[6.2.1.0²,⁴.0⁴,⁷]UNDECANE
CAS RN 167904-65-6

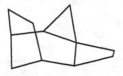

tetracyclo[6.2.1.0²,⁶.0⁵,¹⁰]undecane 1α, 2β, 7α, 8α

ΔH_f^o = - 52.0 kcal/mol

TETRACYCLO[6.2.1.0²,⁶.0⁵,¹⁰]UNDECANE
CAS RN 59014-97-0

2,8-ETHANONORADAMANTANE
6,1,4-ethanylylidene,1H-indene, octahydro

ΔH_f^o = - 12.1 kcal/mol

endo, exo-TETRACYCLO[6.2.1.0^{2,7}.0^{3,5}]UNDECANE
CAS RN   114028-44-3
                  45848-45-1*

2,5-methanocycloprop[a]indene, decahydro 1aα, 1bβ, 2α, 5α, 5aβ, 6aα

ΔH_f^o = 15.0 kcal/mol

---

exo, endo-TETRACYCLO[6.2.1.0^{2,7}.0^{3,5}]UNDECANE
CAS RN   1777-44-2
                  45848-45-1*

2,5-methanocycloprop[a]indene, decahydro 1aα, 1bβ, 2β, 5β, 5aβ, 6aα

ΔH_f^o = 17.3 kcal/mol

TETRACYCLO[6.2.1.0^{2,7}.0^{4,9}]UNDECANE
CAS RN   100762-58-1 (S)   100762-62-7 (±)
                  74867-97-3 (-)   59015-15-5*

[2.1]TRIBLATTANE
METHANOTWISTANE
1,5:2,4-dimethano-1H-indene, octahydro 1α, 2α, 3aβ, 4α, 5α, 7aβ

ΔH_f^o = - 1.4 kcal/mol

TETRACYCLO[6.3.0.0^{2,6}.0^{5,9}]UNDECANE
CAS RN   59015-02-0

1,2,4-[1]-propanyl[3]ylidenepentalene, octahydro
HOMOHYPOSTROPHENE

ΔH_f^o = - 3.3 kcal/mol

TETRACYCLO[8.1.0.0^{1,3}.0^{3,5}]UNDECANE
CAS RN   151587-95-0

ΔH_f^o = 65.0 kcal/mol

TETRACYCLO[8.1.0.0^{1,3}.0^{5,7}]UNDECANE
CAS RN   184759-89-5
               151587-99-4*

tetracyclo[8.1.0.0^{1,3}.0^{5,7}]undecane 3α, 5α, 7α, 10β

ΔH_f^o = 56.4 kcal/mol

TETRACYCLO[8.1.0.0^{1,3}.0^{5,7}]UNDECANE

CAS RN   184759-88-4
            151587-99-4*

tetracyclo[8.1.0.0^{1,3}.0^{5,7}]undecane 3α, 5β, 7β, 10β

ΔH_f^o = 58.7 kcal/mol

TRISPIRO[2.0.2.0.2.2]UNDECANE

CAS RN   33018-17-6

ΔH_f^o = 32.3 kcal/mol

TRISPIRO[2.0.2.1.2.1]UNDECANE

CAS RN   33018-18-7

ΔH_f^o = 35.9 kcal/mol

TRISPIRO[2.0.3.0.3.0]UNDECANE
CAS RN   65656-42-0

ΔH_f^o = 127.3 kcal/mol

TRISPIRO[2.1.1.2.1.1]UNDECANE
CAS RN   21170-76-3

ΔH_f^o = 137.3 kcal/mol

C. PENTACYCLOUNDECANES - $C_{11}H_{14}$

DISPIRO[CYCLOPROPANE-1,3'-TRICYCLO[3.2.0.0^{2,7}]HEPTANE-6',1"-CYCLOPROPANE
CAS RN   94348-13-7

ΔH_f^o = 103.6 kcal/mol

endo-**PENTACYCLO[3.3.3.0^{2,4}.0^{6,8}.0^{9,11}]UNDECANE**
CAS RN   70469-85-5
           27197-30-4*

TRISHOMOBARRELENE

ΔH_f^o = 73.7 kcal/mol

PENTACYCLO[5.3.1.0^{2,4}.0^{2,5}.0^{4,9}]UNDECANE
CAS RN   73586-31-9

2,4-METHANO-2,4-DEHYDROADAMANTANE
2,4:2a,5b-dimethano-2H-cycloprop[cd]indene, hexahydro

ΔH_f^o = 67.2 kcal/mol

123

PENTACYCLO[5.3.1.0^{2,4}.0^{3,6}.0^{5,9}]UNDECANE
CAS RN   58008-63-2

3,5-DEHYDRONORICEANE
2,5-methano-1,6,7-metheno-1H-indene, octahydro

ΔH_f^o = 40.4 kcal/mol

PENTACYCLO[5.3.1.0^{2,5}.0^{3,5}.0^{3,9}]UNDECANE
CAS RN   107616-72-8

2,3-METHANO-2,4-DIDEHYDROADAMANTANE
1,4-methano-3H,6H-cyclopropa[1,3]cycloprop[1,2,3-cd]indene, hexahydro

ΔH_f^o = 64.5 kcal/mol

PENTACYCLO[5.4.0 0^{2,5}.0^{3,9}.0^{4,8}]UNDECANE
CAS RN   62415-12-7
           74867-99-5 (S)

1,3-ETHANOMETHANOCUBANE
[2.1.0]TRIBLATTANE
C$_1$-HOMOBASKETANE
1,2,4-methenocyclobut[cd]indene, decahydro

ΔH_f^o = 38.7 kcal/mol

PENTACYCLO[5.4.0.0^{2,6}.0^{3,10}.0^{5,9}]UNDECANE
CAS RN   4421-32-3

ONE-WINGED BIRD CAGE HYDROCARBON
1,2,4-ethanylylidene-1H-cyclobuta[cd]pentalene, octahydro

ΔH_f^o = 20.7 kcal/mol

PENTACYCLO[5.4.0.0^{2,10}.0^{3,9}.0^{8,11}]UNDECANE
CAS RN   61304-39-0

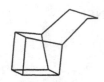

TRISHOMOCUBANE

ΔH_f^o = 81.1 kcal/mol

PENTACYCLO[6.2.1.0^{2,7}.0^{3,5}.0^{4,6}]UNDECANE
CAS RN   115461-01-3

4,7-methano-1,2,3-metheno-1H-indene, octahydro 3aα, 4α, 7α, 7aα

ΔH_f^o = 73.6 kcal/mol

PENTACYCLO[6.3.0.0^{2,6}.0^{3,10}.0^{5,9}]UNDECANE
CAS RN   61473-83-4
                 61473-77-6 (S)
                 30114-56-8*

D$_3$-TRISHOMOCUBANE
TRIHOMOCUBANE
[1.1.1]TRIBLATTANE
1,3,5-methenocyclopenta[cd]pentalene, decahydro

ΔH_f^o = 11.1 kcal/mol

SPIRO[CYCLOPROPANE-1,5-TETRACYCLO[4.3.0.0^{2,4}.0^{3,7}]NONANE]
CAS RN   40756-07-8

spiro[cyclopropane-1,5'(1'H)-[1,2,4]methenopentalene], hexahydro

ΔH_f^o = 51.3 kcal/mol

---

endo, endo-**SPIRO[CYCLOPROPANE-1,9'-TETRACYCLO[3.3.1.0^{2,4}.0^{6,8}]NONANE]**
CAS RN   97056-74-1
                 97558-55-1*

spiro[cyclopropane-1,9'-tetracyclo[3.3.1.0^{2,4}.0$^{6.8}$]nonane], 1'α, 2'β, 4'β, 5'α, 6'β, 8'β

ΔH_f^o = 129.1 kcal/mol

***endo*, *exo*-SPIRO[CYCLOPROPANE-1,9'-TETRACYCLO[3.3.1.0²,⁴.0⁶,⁸]NONANE]**

CAS RN   97101-67-2
              97558-55-1*

spiro[cyclopropane-1,9'-tetracyclo[3.3.1.0²,⁴.0⁶·⁸]nonane] 1'α, 2'α, 4'α, 5'α, 6'β, 8'β

ΔH_f^o = 103.7 kcal/mol

***exo*, *exo*-SPIRO[CYCLOPROPANE-1,9'-TETRACYCLO[3.3.1.0²,⁴.0⁶,⁸]NONANE]**

CAS RN   127759-53-9
              97558-55-1*

spiro[cyclopropane-1,9'-tetracyclo[3.3.1.0²,⁴.0⁶·⁸]nonane], 1'α, 2'α, 4'α, 5'α, 6'α, 8'α

ΔH_f^o = 103.3 kcal/mol

SPIRO[CYCLOPROPANE-1,3'-TETRACYCLO[3.3.1.0²,⁸.0⁴,⁶]NONANE]

CAS RN   67654-18-6

TRIASTERANE-3-SPIROCYCLOPROPANE

ΔH_f^o = 36.0 kcal/mol

---

cis-TETRASPIRO[2.0.0.0.2.1.1.1]UNDECANE
CAS RN   129872-36-2
                     130354-20-2*

cis-[5]TRIANGULANE

ΔH_f^o = 136.7 kcal/mol

---

trans-TETRASPIRO[2.0.0.0.2.1.1.1]UNDECANE
CAS RN   129872-30-6
                     130354-20-2*

trans-[5]TRIANGULANE

ΔH_f^o = 134.4 kcal/mol

TETRASPIRO[2.0.0.2.0.2.0.1]UNDECANE
CAS RN   136028-45-0

ΔH_f^o = 134.6 kcal/mol

D. HEXACYCLOUNDECANES - $C_{11}H_{12}$

HEXACYCLO[5.4.0.0^{1,8}.0^{2,4}.0^{3,5}.0^{7,9}]UNDECANE
CAS RN   73320-82-8

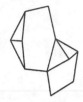

1,3a,7a:4,5,6-dimetheno-1H-indene, hexahydro

ΔH_f^o = 141.0 kcal/mol

HEXACYCLO[5.4.0.0^{2,5}.0^{3,8}.0^{4,6}.0^{9,11}]UNDECANE
CAS RN   67328-36-3

1,2,4-methenodicyclopropa[a,e]pentalene, decahydro 1α, 1aβ, 1bβ, 2α, 2aα, 3aα, 3bβ, 4α, 4aβ

ΔH_f^o = 91.4 kcal/mol

HEXACYCLO[5.4.0.0^{2,5}.0^{3,10}.0^{4,8}.0^{9,11}]UNDECANE
CAS RN   58229-25-7

1,2,4-methenocyclopenta[cd]cyclopropa[gh]pentalene,decahydro

ΔH_f^o = 59.8 kcal/mol

HEXACYCLO[5.4.0.0^{2,6}.0^{3,10}.0^{5,9}.0^{8,11}]UNDECANE
CAS RN   25107-14-6

CHURCHANE
HOMOPENTAPRISMANE
1,2,3,5-ethanediylidene-1H-cyclobuta[cd]pentalene, octahydro

ΔH_f^o = 72.2 kcal/mol

HEXACYCLO[5.4.0.0^{2,11}.0^{3,5}.0^{4,9}.0^{6,8}]UNDECANE
CAS RN   62412-03-7

1,6-*HOMODIADEMANE*
1,2-methanotricycloprop[cd,f,hi]indene, decahydro

ΔH_f^o = 57.7 kcal/mol

SPIRO[CYCLOPROPANE-1,9'-PENTACYCLO[4.3.0.0^{2,4}.0^{3,8}.0^{5,7}]NONANE]
CAS RN   65915-88-0

spiro[cyclopropane-1,3'-[1,2]methanodicyclopropa[cd,gh]pentalene], octahydro

ΔH_f^o = 104.0 kcal/mol

8
POLYCYCLODODECANES
A. TRICYCLODODECANES - $C_{12}H_{20}$

DISPIRO[2.0.2.6]DODECANE
CAS RN   64601-40-7

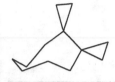

ΔH_f^o = 4.1 kcal/mol

DISPIRO[4.1.4.1]DODECANE
CAS RN   185-01-3

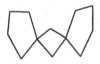

ΔH_f^o = - 10.2 kcal/mol

SPIRO[BICYCLO[2.2.1]HEPTANE-2,1'-CYCLOHEXANE]
CAS RN   172-93-0

1,4-endo-METHYLENESPIRO[5.5]UNDECANE
SPIRO[CYCLOHEXANE-1,2'-NORBORNANE]

ΔH_f^o = - 29.3 kcal/mol

SPIRO[BICYCLO[4.1.0]HEPTANE-7,1'-CYCLOHEXANE]
CAS RN   181-14-6

SPIRO[CYCLOHEXANE-1,7' NORCARANE]
1,1-PENTAMETHYLENE-2,3-TETRAMETHYLENECYCLOPROPANE

ΔH_f^o = - 16.5 kcal/mol

TRICYCLO[4.2.2.2^{2,5}]DODECANE
CAS RN   259-93-8

PERHYDRO[0.0]PARACYCLOPHANE
PERHYDRO[0.0](1,4)-CYCLOPHANE

ΔH_f^o = - 12.3 kcal/mol

TRICYCLO[4.3.3.0^{1,6}]DODECANE
CAS RN   7161-28-6

[4.3.3]PROPELLANE
3a,7a-propano-1H-indene, hexahydro

ΔH_f^o = - 32.3 kcal/mol

TRICYCLO[4.3.3.0^{2,5}]DODECANE
CAS RN   19566-14-4

ΔH_f^o = - 1.9 kcal/mol

TRICYCLO[4.4.1.1^{3,8}]DODECANE
CAS RN   35847-47-3

1,5-BISHOMOADAMANTANE

ΔH_f^o = - 22.8 kcal/mol

TRICYCLO[4.4.1.1^{3,9}]DODECANE
CAS RN   36071-59-7

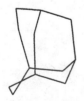

1,3-BISHOMOADAMANTANE

ΔH_f^o = - 23.6 kcal/mol

TRICYCLO[4.4.2.0^{1,6}]DODECANE
CAS RN   7620-88-4

[4.4.2]PROPELLANE
4a,8a-ethanonaphthalene, octahydro

ΔH_f^o = - 15.7 kcal/mol

TRICYCLO[4.4.2.0^{2,5}]DODECANE
CAS RN   5020-34-8

ΔH_f^o = 1.3 kcal/mol

TRICYCLO[5.3.1.1^{3,9}]DODECANE
CAS RN   36071-50-8

1,1-BISHOMOADAMANTANE

ΔH_f^o = - 23.6 kcal/mol

---

cis-TRICYCLO[5.3.2.0^{1,6}]DODECANE
CAS RN 62859-77-2

2H-1,4a-ethanonaphthalene, octahydro 1α, 4aα, 8aα

ΔH_f^o = - 33.6 kcal/mol

trans-TRICYCLO[5.3.2.0^{1,6}]DODECANE
CAS RN 62797-91-5

2H-1,4a-ethanonaphthalene, octahydro 1α, 4aα, 8aβ

ΔH_f^o = - 30.8 kcal/mol

TRICYCLO[5.3.2.0^{1,7}]DODECANE
CAS RN 43043-81-8

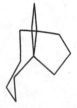

[5.3.2]PROPELLANE
1H,4H-3a,8a-ethanoazulene, hexahydro

ΔH_f^o = - 13.1 kcal/mol

TRICYCLO[5.4.1.0^{1,5}]DODECANE
CAS RN   62859-79-4
                6538-27-8*

3a,7-methano-3aH-cyclopentacyclooctene, decahydro 3aα, 7α, 9aα

ΔH_f^o = - 34.4 kcal/mol

TRICYCLO[5.4.1.0^{1,5}]DODECANE
CAS RN   62859-78-3
                6538-27-8*

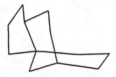

3a,7-methano-3aH-cyclopentacyclooctene, decahydro 3aα, 7α, 9aβ

ΔH_f^o = - 31.9 kcal/mol

TRICYCLO[5.4.1.0^{1,7}]DODECANE
CAS RN   152484-03-2

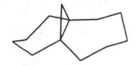

[5.4.1]PROPELLANE
4a,9a-methano-5H-benzocycloheptene, octahydro

ΔH_f^o = - 19.3 kcal/mol

TRICYCLO[5.4.1.0^{4,12}]DODECANE
CAS RN   826-68-6

cyclopent[cd]azulene, dodecahydro

ΔH_f^o = - 30.9 kcal/mol

TRICYCLO[5.5.0.0^{2,8}]DODECANE
CAS RN   61506-29-4

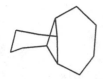

ΔH_f^o = - 11.4 kcal/mol

---

cis-TRICYCLO[6.2.2.0^{2,7}]DODECANE
CAS RN   5216-90-0
                 703-34-4*

cis-1,4-ethanonaphthalene, decahydro

ΔH_f^o = - 31.2 kcal/mol

---

trans-TRICYCLO[6.2.2.0^{2,7}]DODECANE
CAS RN   5216-91-1
            703-34-4*

trans-1,4-ethanonaphthalene, decahydro

ΔH_f^o = - 31.3 kcal/mol

TRICYCLO[6.2.2.0^{4,9}]DODECANE
 CAS RN   54516-68-8*   51705-38-5
                51705-39-6   51705-40-4
                51644-47-4

1,7-ethanonaphthalene, decahydro

ΔH_f^o = - 36.9 kcal/mol

---

cis-TRICYCLO[6.3.1.0^{1,6}]DODECANE
CAS RN   62797-90-4
            70878-80-7*

4a,8-methano-4aH-benzocycloheptene, decahydro 4aα, 8α, 9aα

ΔH_f^o = -32.6 kcal/mol

***trans*-TRICYCLO[6.3.1.0^{1,6}]DODECANE** ·
CAS RN   62859-76-1
          70878-80-7*

4a,8-methano-4aH-benzocycloheptene, decahydro 4aα, 8α, 9aβ

ΔH_f^o = - 34.9 kcal/mol

TRICYCLO[6.3.1.0^{2,6}]DODECANE
CAS RN   62859-70-5
          75828-04-5*

4,8-methano-1H-cyclopentacyclooctene, decahydro 3aα, 4α, 8α, 9aα

ΔH_f^o = -28.7 kcal/mol

TRICYCLO[6.3.1.0^{2,6}]DODECANE
CAS RN   62861-04-5
          75828-04-5*

4,8-methano-1H-cyclopentacyclooctene, decahydro 3aα, 4α, 8α, 9aβ

ΔH_f^o = - 30.3 kcal/mol

TRICYCLO[6.3.1.0^{2,6}]DODECANE
CAS RN   62859-71-6
            75828-04-5*

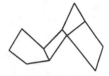

4,8-methano-1H-cyclopentacyclooctene, decahydro 3aα, 4β, 8β, 9aα

ΔH_f^o = - 32.6 kcal/mol

TRICYCLO[6.3.1.0^{2,6}]DODECANE
CAS RN   62797-87-9
            75828-04-5*

4,8-methano-1H-cyclopentacyclooctene, decahydro 3aα, 4β, 8β, 9aβ

ΔH_f^o = - 29.2 kcal/mol

---

cis, cis, cis-**TRICYCLO[6.3.1.0^{4,12}]DODECANE**
CAS RN   38113-44-9
            2146-36-3*

cis, cis, cis-UFOLANE
cis, cis, cis-PERHYDROACENAPHTHENE
acenaphthylene, dodecahydro 2aα, 5aα, 8aα, 8bα

ΔH_f^o = - 33.0 kcal/mol

***cis, cis, trans*-TRICYCLO[6.3.1.0⁴,¹²]DODECANE**

CAS RN   38113-42-7(±)
                2146-36-3*

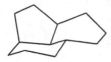

cis, cis, trans-UFOLANE
cis, cis, trans-PERHYDROACENAPHTHENE
acenaphthylene, dodecahydro 2aα, 5aα, 8aβ, 8bα

ΔH_f^o = - 34.9 kcal/mol

***cis, trans, cis*-TRICYCLO[6.3.1.0⁴,¹²]DODECANE**

CAS RN   38113-41-6
                2146-36-3*

cis, trans, cis-UFOLANE
cis, trans, cis-PERHYDROACENAPHTHENE
acenaphthylene, dodecahydro 2aα, 5aβ, 8aα, 8bα

ΔH_f^o = - 36.6 kcal/mol

***cis, trans, trans*-TRICYCLO[6.3.1.0⁴,¹²]DODECANE**

CAS RN   38113-40-5 (±)
                2146-36-3*

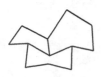

cis, trans, trans-UFOLANE
cis, trans, trans-PERHYDOACENAPHTHENE
acenaphthylene, dodecahydro 2aα, 5aα, 8aβ, 8bβ

ΔH_f^o = - 34.3 kcal/mol

***trans, cis, trans*-TRICYCLO[6.3.1.0^{4,12}]DODECANE**
CAS RN   19222-42-5
                2146-36-3*

trans, cis, trans-UFOLANE
trans, cis, trans-PERHYDROACENAPHTHENE
acenaphthylene, dodecahydro 2aα, 5aβ, 8aα, 8bβ

ΔH_f^o = - 27.6 kcal/mol

***trans, trans, trans*-TRICYCLO[6.3.1.0^{4,12}]DODECANE**
CAS RN   38113-39-2
                2146-36-3*

trans, trans, trans-UFOLANE
trans, trans, trans-PERHYDROACENAPHTHENE
acenaphthylene, dodecahydro 2aα, 5aα, 8aα, 8bβ

ΔH_f^o = - 37.8 kcal/mol

TRICYCLO[6.4.0.0^{1,5}]DODECANE
CAS RN   120052-84-8 (±)

1H-cyclopent[c]indene, decahydro, 3aα, 5aα

ΔH_f^o = - 32.1 kcal/mol

TRICYCLO[6.4.0.0^{1,5}]DODECANE
CAS RN   119972-33-7 (±)

1H-cyclopent[c]indene, decahydro, 3aα, 5aβ

ΔH_f^o = - 4.7 kcal/mol

---

cis-cisoid-cis-TRICYCLO[6.4.0.0^{2,6}]DODECANE
CAS RN   62859-69-2
            51302-75-1*

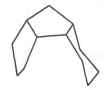

cyclopent[a]indene, dodecahydro 3aα, 3bα, 7aα, 8aα

ΔH_f^o = - 33.0 kcal/mol

---

cis-cisoid-trans-TRICYCLO[6.4.0.0^{2,6}]DODECANE
CAS RN   100761-64-6
            51302-75-1*

cyclopent[a]indene, dodecahydro 3aα, 3bα, 7aβ, 8aα

ΔH_f^o = - 24.7 kcal/mol

***cis-transoid-cis*-TRICYCLO[6.4.0.0²,⁶]DODECANE**

CAS RN   62859-68-1
                 51302-75-1*

cyclopent[a]indene, dodecahydro 3aα, 3bβ, 7aβ, 8aα

ΔH_f^o = - 34.9 kcal/mol

***trans-cisoid-trans*-TRICYCLO[6.4.0.0²,⁶]DODECANE**

CAS RN   62859-67-0
                 51302-75-1*

cyclopent[a]indene, dodecahydro 3aα, 3bα, 7aβ, 8aβ

ΔH_f^o = - 20.5 kcal/mol

***trans-transoid-cis*-TRICYCLO[6.4.0.0²,⁶]DODECANE**

CAS RN   100761-63-5
                 51302-75-1*

cyclopent[a]indene, dodecahydro 3aα, 3bβ, 7aα, 8aα

ΔH_f^o = - 36.0 kcal/mol

***trans-transoid-trans*-TRICYCLO[6.4.0.0^{2,6}]DODECANE**
CAS RN   62859-66-9
           51302-75-1*

cyclopent[a]indene, dodecahydro 3aα, 3bβ, 7aα, 8aβ

ΔH_f^o = - 26.3 kcal/mol

***cis-cisoid-cis*-TRICYCLO[6.4.0.0^{2,7}]DODECANE**
CAS RN   29782-48-7
           53485-49-7*

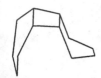

biphenylene, dodecahydro 4aα, 4bα, 8aα, 8bα

ΔH_f^o = - 6.8 kcal/mol

***cis-cisoid-trans*-TRICYCLO[6.4.0.0^{2,7}]DODECANE**
CAS RN   51319-08-5
           53485-49-7*

biphenylene, dodecahydro 4aα, 4bα, 8aα, 8bβ

ΔH_f^o = - 7.0 kcal/mol

---

cis-transoid-cis-TRICYCLO[6.4.0.0^{2,7}]DODECANE
CAS RN   29782-49-8
               53485-49-7*

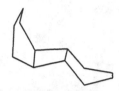

biphenylene, dodecahydro 4aα, 4bβ, 8aβ, 8bα

$\Delta H_f^o = -13.6$ kcal/mol

---

trans-cisoid-trans-TRICYCLO[6.4.0.0^{2,7}]DODECANE
CAS RN   939-02-6
               53485-49-7*

biphenylene, dodecahydro 4aα, 4bα, 8aβ, 8bβ

$\Delta H_f^o = -10.4$ kcal/mol

---

trans-transoid-trans-TRICYCLO[6.4.0.0^{2,7}]DODECANE
CAS RN   51319-07-4

biphenylene, dodecahydro 4aα, 4bβ, 8aα, 8bβ

$\Delta H_f^o = -6.6$ kcal/mol

TRICYCLO[6.4.0.0²,¹⁰]DODECANE
CAS RN   56670-27-0

1,4-methano-1H-cyclopentacyclooctene, decahydro

ΔH_f^o = -17.5 kcal/mol

TRICYCLO[6.4.0.0⁴,⁹]DODECANE
CAS RN   71274-72-1   68122-48-5 (S)
           68122-49-6 (R)   99651-80-6 (±)

1,5-ethanonaphthalene, decahydro

ΔH_f^o = -32.4 kcal/mol

TRICYCLO[7.2.1.0¹,⁶]DODECANE
CAS RN   62859-75-0
           3403-70-3*

4a,7-methano-4aH-benzocycloheptene, decahydro 4aα, 7α, 9aα

ΔH_f^o = - 36.0 kcal/mol

TRICYCLO[7.2.1.0^{1,6}]DODECANE
CAS RN   62797-89-1
               34303-70-3*

4a,7-methano-4aH-benzocycloheptene, decahydro 4aα, 7α, 9aβ

ΔH_f^o = - 37.2 kcal/mol

TRICYCLO[7.2.1.0^{2,7}]DODECANE
CAS RN   62859-74-9
               75828-03-4*

5,8-methano-1H-benzocycloheptene, decahydro 4aα, 5α, 8α, 9aα

ΔH_f^o = - 29.4 kcal/mol

TRICYCLO[7.2.1.0^{2,7}]DODECANE
CAS RN   62797-88-0
               75828-03-4*

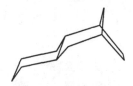

5,8-methano-1H-benzocycloheptene, decahydro 4aα, 5α, 8α, 9aβ

ΔH_f^o = - 36.3 kcal/mol

TRICYCLO[7.2.1.0^{2,7}]DODECANE
CAS RN   62859-73-8
            75828-03-4*

5,8-methano-1H-benzocycloheptene, decahydro 4aα, 5β, 8β, 9aα

ΔH_f^o = - 33.9 kcal/mol

---

cis, trans-TRICYCLO[7.3.0.0^{1,5}]DODECANE
CAS RN   93887-91-3

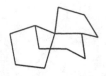

1H-cyclopent[d]indene, decahydro, 3aα, 6aα

ΔH_f^o = - 30.6 kcal/mol

---

cis-cisoid-cis-TRICYCLO[7.3.0.0^{2,6}]DODECANE
CAS RN   30159-14-9
            30767-91-0*

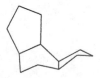

as-indacene, dodecahydro 3aα, 5aα, 8aα, 8bα

ΔH_f^o = - 30.2 kcal/mol

---

cis-cisoid-trans-TRICYCLO[7.3.0.0^{2,6}]DODECANE

CAS RN   30159-13-8
               30767-91-0*

as-indacene, dodecahydro 3aα, 5aβ, 8aα, 8bα

ΔH_f^o = - 31.9 kcal/mol

---

cis-transoid-cis-TRICYCLO[7.3.0.0^{2,6}]DODECANE

CAS RN   30159-15-0
               30767-91-0*

as-indacene, dodecahydro 3aα, 5aβ, 8aβ, 8bα

ΔH_f^o = - 32.4 kcal/mol

---

cis-transoid-trans-TRICYCLO[7.3.0.0^{2,6}]DODECANE

CAS RN   30159-16-1
               30767-91-0*

as-indacene, dodecahydro 3aα, 5aα, 8aα, 8bβ

ΔH_f^o = - 32.6 kcal/mol

***trans-cisoid-trans*-TRICYCLO[7.3.0.0^{2,6}]DODECANE**
CAS RN   30159-12-7
               30767-91-0*

as-indacene, dodecahydro 3aα, 5aα, 8aβ, 8bβ

ΔH_f^o = - 24.8 kcal/mol

***trans-transoid-trans*-TRICYCLO[7.3.0.0^{2,6}]DODECANE**
CAS RN   30159-17-2
               30767-91-0*

as-indacene, dodecahydro 3aα, 5aβ, 8aα, 8bβ

ΔH_f^o = - 33.3 kcal/mol

TRICYCLO[7.3.0.0^{3,7}]DODECANE
CAS RN   54158-72-4

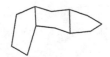

s-indacene, dodecahydro

ΔH_f^o = - 32.7 kcal/mol

cis, trans-**TRICYCLO[9.1.0.0⁴,⁶]DODECANE**

cis, trans-**TRICYCLO[9.1.0.0^{4,6}]DODECANE**

CAS RN   24316-03-8

ΔH_f^o = 13.0 kcal/mol

B. TETRACYCLODODECANES - $C_{12}H_{18}$

DISPIRO[CYCLOPENTANE-1,2'-BICYCLO[1.1.0]BUTANE-4',1''-CYCLOPENTANE]
CAS RN   114310-66-6

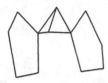

ΔH_f^o = 48.3 kcal/mol

DISPIRO[CYCLOPROPANE-1,2'-BICYCLO[2.2.2]OCTANE-3',1''-CYCLOPROPANE]
CAS RN   40827-30-3

ΔH_f^o = 12.7 kcal/mol

SPIRO[BICYCLO[4.1.0]HEPTANE-7,6'-BICYCLO[3.1.0]HEXANE]
CAS RN   56424-64-7

ΔH_f^o = 33.2 kcal/mol

153

SPIRO[CYCLOPROPANE-1,2'-TRICYCLO[3.3.1.1^{3,7}]DECANE]
CAS RN   19740-32-0

CYCLOPROPANOADAMANTANE
SPIRO[ADAMANTANE-2,1'-CYCLOPROPANE]

ΔH_f^o = - 12.7 kcal/mol

TETRACYCLO[4.2.2.2^{2,5}.0^{1,6}]DODECANE
CAS RN   77422-57-2

2,5-ETHANO[4.2.2]PROPELLANE

ΔH_f^o = 24.5 kcal/mol

TETRACYCLO[5.3.1.1^{2,6}.0^{4,9}]DODECANE
CAS RN   53283-19-5

ICEANE
WURTZITANE
2,7:3,6-dimethanonaphthalene, decahydro

ΔH_f^o = - 19.0 kcal/mol

TETRACYCLO[5.4.1.0^{1,7}.0^{4,12}]DODECANE
CAS RN 54008-05-8

benzo[1,3]cyclopropa[1,2,3-cd]pentalene, decahydro

ΔH_f^o = - 0.8 kcal/mol

***endo, endo*-TETRACYCLO[5.4.1.0^{2,6}.0^{8,11}]DODECANE**
CAS RN 92935-04-1

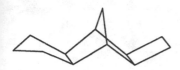

3,7-methano-1H-cyclobut[f]indene, decahydro 2aα, 3β, 3aα, 6aα, 7β, 7aα

ΔH_f^o = 5.6 kcal/mol

***endo, exo*-TETRACYCLO[5.4.1.0^{2,6}.0^{8,11}]DODECANE**
CAS RN 92807-91-5

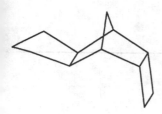

3,7-methano-1H-cyclobut[f]indene, decahydro 2aα, 3β, 3aβ, 6aβ, 7β, 7aα

ΔH_f^o = 10.2 kcal/mol

TETRACYCLO[5.4.1.0³,⁵.0⁴,¹²]DODECANE
CAS RN   54008-06-9

cyclohepta[cd]cyclopropa[gh]pentalene, decahydro

ΔH_f^o = 42.3 kcal/mol

TETRACYCLO[5.4.1.0⁴,¹².0⁹,¹²]DODECANE
CAS RN   82353-32-0 (R)

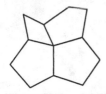

all-cis-[4.5.5.5]FENESTRANE
1,8-methanocyclopenta[c]pentalene, decahydro 1α, 3aα, 5aβ, 8β

ΔH_f^o = - 1.6 kcal/mol

TETRACYCLO[6.2.1.1²,⁵.0¹,⁶]DODECANE
CAS RN   83134-52-5

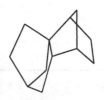

2H-1,4:4a,7-dimethanonaphthalene, octahydro 1α, 4α, 4aβ, 7β, 8aα

ΔH_f^o = - 3.1 kcal/mol

---

endo, endo-TETRACYCLO[6.2.1.1^{3,6}.0^{2,7}]DODECANE
CAS RN   58865-53-5
                32021-58-2*

DIMETHANODECALIN
1,4:5,8-dimethanonaphthalene, decahydro 1α, 4α, 4aβ, 5α, 8α, 8aβ
ΔH_f^o = 6.6 kcal/mol

---

exo, endo-TETRACYCLO[6.2.1.1^{3,6}.0^{2,7}]DODECANE
CAS RN   15914-95-1
                32021-58-2*

DIMETHANODECALIN
1,4:5,8-dimethanonaphthalene, decahydro 1α, 4α, 4aα, 5β, 8β, 8aα
ΔH_f^o = - 2.5 kcal/mol

---

exo, exo-TETRACYCLO[6.2.1.1^{3,6}.0^{2,7}]DODECANE
CAS RN   53862-33-2
                32021-58-2*

DIMETHANODECALIN
1,4:5,8-dimethanonaphthalene, decahydro 1α, 4α, 4aα, 5α, 8α, 8aα
ΔH_f^o = - 3.7 kcal/mol

TETRACYCLO[6.2.2.0^{2,7}.0^{4,9}]DODECANE
CAS RN   54445-71-5
             74843-75-7 ($\pm$)
             74867-98-4 (R)

DITWISTANE
[8]DITWISTANE
[2.2]TRIBLATTANE
1,6:2,5-dimethanonaphthalene, decahydro

ΔH_f^o = - 14.0 kcal/mol

TETRACYCLO[6.3.1.0^{2,6}.0^{5,10}]DODECANE
CAS RN   15002-90-1

2,4-ETHANOADAMANTANE
4,1,6-[1,2,3]propanetriyl-1H-indene, octahydro

ΔH_f^o = - 24.3 kcal/mol

TETRACYCLO[6.3.1.0^{2,7}.0^{3,5}]DODECANE
CAS RN   82072-01-3

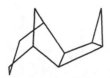

2,6-methano-1H-cycloprop[a]azulene, decahydro 1aα, 1bβ, 2β, 6β, 6aβ, 7aα

ΔH_f^o = 5.7 kcal/mol

TETRACYCLO[6.4.0.0^{2,7}.0^{4,11}]DODECANE
CAS RN   40826-92-4

ΔH_f^o = 6.0 kcal/mol

TETRACYCLO[6.4.0.0^{2,10}.0^{5,9}]DODECANE
CAS RN   53438-64-5

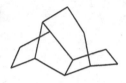

1,4,7-metheno-1H-cyclopentacyclooctene, decahydro

ΔH_f^o = - 12.2 kcal/mol

TETRACYCLO[6.4.0.0^{2,11}.0^{3,10}]DODECANE
CAS RN   100693-73-0
            100762-60-5 (±)

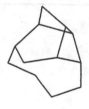

[4.0]TRIBLATTANE

ΔH_f^o = 26.2 kcal/mol

TETRACYCLO[6.4.0.0⁴,¹².0⁵,⁹]DODECANE

CAS RN   18326-54-0

4,1,5-[1]propanyl[3]ylidene-1H-indene, octahydro

ΔH_f^o = - 4.5 kcal/mol

TETRACYCLO[7.2.1.0¹,⁶.0⁸,¹⁰]DODECANE

CAS RN   68276-49-3
            93519-50-7 (±)

NORISHWARANE
1,2a-methano-2aH-cyclopropa[b]naphthalene, decahydro

ΔH_f^o = - 5.2 kcal/mol

TETRACYCLO[7.2.1.0²,⁸.0⁴,¹⁰]DODECANE

CAS RN   100762-63-8 (±)
            100693-74-1 (S)

[3.1]TRIBLATTANE
1,5:2,4-dimethanoazulene, decahydro 1α, 2α, 3aβ, 4α, 5α, 8aβ

ΔH_f^o = - 1.9 kcal/mol

---

all-cis-TETRACYCLO[7.2.1.0^{4,11}.0^{6,10}]DODECANE
CAS RN   60606-97-5

TETRAQUINANE
dicyclopenta[cd,gh]pentalene, dodecahydro

ΔH_f^o = - 18.7 kcal/mol

TETRACYCLO[8.2.0.0^{2,5}.0^{6,9}]DODECANE
CAS RN   87422-12-6

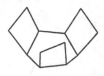

cis-tris-[2.2.2]-σ-HOMOBENZENE
tetracyclo[8.2.0.0^{2,5}.0^{6,9}]dodecane 1α, 2α, 5α, 6α, 9α, 10α

ΔH_f^o = 48.6 kcal/mol

TETRACYCLO[9.1.0.0^{1,3}.0^{6,8}]DODECANE
CAS RN   184643-58-1

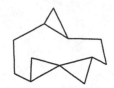

ΔH_f^o = 51.9 kcal/mol

TETRACYCLO[9.1.0.0^{3,5}.0^{7,9}]DODECANE
CAS RN   285-55-2*

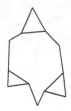

all cis–HEXAHOMOBENZENE

ΔH_f^o = 41.0 kcal/mol

TRISPIRO[2.0.2.1.2.2]DODECANE
CAS RN   116316-71-3

ΔH_f^o = 26.6 kcal/mol

TRISPIRO[3.0.3.0.3.0]DODECANE
CAS RN   126379-27-9

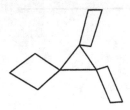

[3.4]ROTANE

ΔH_f^o = 132.9 kcal/mol

C. PENTACYCLODODECANES - $C_{12}H_{16}$

***trans*-DISPIRO[CYCLOBUTANE-1,3'-TRICYCLO[3.2.1.0^{2,4}]HEXANE-6',1''-CYCLOBUTANE]**
CAS RN   114640-69-6

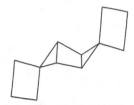

ΔH_f^o = 98.5 kcal/mol

PENTACYCLO[4.4.2.0^{2,10}.0^{3,5}.0^{7,9}]DODECANE
CAS RN   82569-90-2
              28648-36-4*

exo,exo-DIHYDROBISHOMOBULLVALENE
pentacyclo[4.4.2.^{2,10}.0^{3,5}.0^{7,9}]dodecane 1α, 2β, 4β, 5β, 6α, 7β, 8β, 10β

ΔH_f^o = 80.5 kcal/mol

PENTACYCLO[5.3.1.1^{2,6}.0^{1,7}.0^{2,6}]DODECANE
CAS RN   85739-36-2

anti-BIS[3.2.1]PROPELLANE
dimethanocyclobuta[1,2:3,4]dicyclopentene, tetrahydro 3aα, 3bβ, 6aβ, 6bα

ΔH_f^o = 52.2 kcal/mol

PENTACYCLO[5.4.1.0^{3,10}.0^{4,12}.0^{5,9}]DODECANE
CAS RN   34033-63-1

2,5,7-metheno-1H-cyclopenta[a]pentalene, decahydro

ΔH_f^o = 15.1 kcal/mol

PENTACYCLO[5.4.1.0^{3,10}.0^{5,9}.0^{8,11}]DODECANE
CAS RN   85850-64-2

[4]PERISTYLANE
1,6:3,4-dimethanocyclobuta[1,2:3,4]dicyclopentene, decahydro

ΔH_f^o = 13.9 kcal/mol

PENTACYCLO[5.5.0.0^{2,5}.0^{3,9}.0^{4,8}]DODECANE
CAS RN   94319-66-1 (R)

[3.1.0]TRIBLATTANE
1,2,4-metheno-1H-cyclobut[cd]azulene, decahydro 1α, 1aβ, 2α, 3aβ, 4α, 7aβ, 7bβ

ΔH_f^o = 38.6 kcal/mol

PENTACYCLO[6.2.1.1^{2,5}.0^{1,6}.0^{6,10}]DODECANE
CAS RN   94596-71-1

2,3b:4,7-dimethano-3bH-cyclopenta[1,3]cyclopropa[1,2]benzene, octahydro

ΔH_f^o = 26.8 kcal/mol

PENTACYCLO[6.3.1.0^{2,4}.0^{3,7}.0^{5,10}]DODECANE
CAS RN   108643-89-6

3,12-CYCLOICEANE
2,5:3,4-dimethanocyclopropa[de]naphthalene, decahydro

ΔH_f^o = 19.2 kcal/mol

exo, exo-PENTACYCLO[6.3.1.0^{2,4}.0^{5,12}.0^{9,11}]DODECANE
CAS RN   55756-74-6

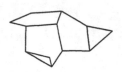

cyclopenta[cd]dicyclopropa[a,f]pentalene, dodecahydro 2aα,2bβ, 3aβ, 3bα, 3cβ, 4aβ, 4bα, 4cα

ΔH_f^o = 64.9 kcal/mol

PENTACYCLO[6.3.1.0²,⁷.0³,⁵.0⁹,¹¹]DODECANE
CAS RN   82110-70-1
          6049-82-7*

2,4-methano-1H-dicyclopropa[a,f]indene, dodecahydro 1aα, 1bβ, 2β, 2aα, 3aα, 4β,4aβ, 5aα

ΔH_f^o = 65.1 kcal/mol

PENTACYCLO[6.3.1.0²,⁷.0³,⁵.0⁹,¹¹]DODECANE
CAS RN   114028-43-2
          6049-82-7*

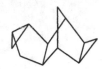

2,4-methano-1H-dicyclopropa[a,f]indene, decahydro 1aα, 1bβ, 2α, 2aβ, 3aβ, 4α, 4aβ, 5aα

ΔH_f^o = 44.8 kcal/mol

PENTACYCLO[6.4.0.0²,⁵.0³,¹⁰.0⁴,⁹]DODECANE
CAS RN   74868-00-1 (-)
          62415-14-9*

1,3-BISETHANOCUBANE
ANSARANE
[2.2.0]TRIBLATTANE
3,10-DEHYDRODITWISTANE
1,2,5-metheno-1H-cyclobuta[de]naphthalene, decahydro 1α, 1aβ, 2α, 4aβ, 5α, 7aβ, 7bβ

ΔH_f^o = 26.9 kcal/mol

PENTACYCLO[6.4.0.0^{2,6}.0^{3,10}.0^{5,9}]DODECANE
CAS RN   150950-17-7   116347-39-8 (R)
          70267-03-7 (±)   70209-47-1 (S)

BISMETHANOTWISTANE
ETHANODITWISTANE
[2.1.1]TRIBLATTANE
1,3,5-metheno-1H-cyclopent[cd]indene, decahydro

ΔH_f^o = 5.2 kcal/mol

PENTACYCLO[6.4.0.0^{2,6}.0^{3,11}.0^{5,9}]DODECANE
CAS RN   150858-10-9

1,3,5-ethanylylidenecyclopenta[cd]pentalene, decahydro

ΔH_f^o = - 3.1 kcal/mol

PENTACYCLO[6.4.0.0^{2,7}.0^{3,11}.0^{6,10}]DODECANE
CAS RN   4421-33-4

1,3,6-ethanylylidenecyclobut[cd]indene, decahydro

ΔH_f^o = 15.1 kcal/mol

PENTACYCLO[6.4.0.0^{2,7}.0^{3,12}.0^{4,11}]DODECANE
CAS RN   67277-98-9

dicyclobuta[def,jkl]biphenylene, dodecahydro

$\Delta H_f^o = 53.6$ kcal/mol

PENTACYCLO[6.4.0.0^{2,7}.0^{4,11}.0^{5,10}]DODECANE
CAS RN   259-77-8

TETRAASTERANE

$\Delta H_f^o = 21.1$ kcal/mol

PENTACYCLO[6.4.0.0^{2,10}.0^{3,7}.0^{4,9}]DODECANE
CAS RN   53283-02-6

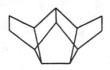

3,4,7-metheno-1H-cyclopenta[a]pentalene, decahydro

$\Delta H_f^o = 10.8$ kcal/mol

PENTACYCLO[6.4.0.0^{2,10}.0^{3,7}.0^{5,9}]DODECANE
CAS RN   6673-99-0

2,4,7-metheno-1H-cyclopenta[a]pentalene, decahydro
1,4-methano-2,5,8-methenonaphthalene, decahydro

ΔH_f^O = 7.9 kcal/mol

PENTACYCLO[6.4.0.0^{2,12}.0^{3,7}.0^{4,11}]DODECANE
CAS RN   67277-99-0

1,2,3-[1]propanyl[3]ylidene-1H-cycloprop[cd]indene, octahydro

ΔH_f^O = 23.6 kcal/mol

PENTACYCLO[7.2.1.0^{2,7}.0^{3,5}.0^{4,8}]DODECANE
CAS RN   73157-58-1

5,8-methano-1,2,4-methenoazulene, decahydro 1α, 2α, 3aβ, 4α, 5α, 8α, 8aβ

ΔH_f^O = 23.4 kcal/mol

---

trans-PENTACYCLO[9.1.0.0^{1,3}.0^{3,5}.0^{5,7}]DODECANE

CAS RN   159618-77-6

ΔH_f^o = 106.0 kcal/mol

PENTACYCLO[9.1.0.0^{1,3}.0^{4,6}.0^{6,8}]DODECANE

CAS RN   151588-01-0

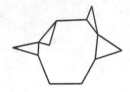

ΔH_f^o = 86.2 kcal/mol

PENTACYCLO[9.1.0.0^{1,3}.0^{4,6}.0^{7,9}]DODECANE

CAS RN   184643-57-0

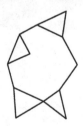

ΔH_f^o = 75.9 kcal/mol

PENTACYCLO[9.1.0.0^{1,3}.0^{5,7}.0^{7,9}]DODECANE
CAS RN   151588-00-0

ΔH_f^o = 95.1 kcal/mol

---

cis, cis, trans-**PENTACYCLO[9.1.0.0^{2,4}.0^{5,7}.0^{8,10}]DODECANE**
CAS RN   105662-05-3

pentacyclo[9.1.0.0^{2,4}.0^{5,7}.0^{8,10}] dodecane 1α, 2α, 4α, 5α, 7α, 8β, 10β, 11α

ΔH_f^o = 64.3 kcal/mol

---

cis, trans, cis-**PENTACYCLO[9.1.0.0^{2,4}.0^{5,7}.0^{8,10}]DODECANE**
CAS RN   105662-06-4

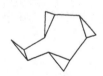

pentacyclo[9.1.0.0^{2,4}.0^{5,7}.0^{8,10}] dodecane 1α, 2α, 4α, 5β, 7β, 8β, 10β, 11α

ΔH_f^o = 57.1 kcal/mol

---

trans, trans, trans-**PENTACYCLO[9.1.0.0^{2,4}.0^{5,7}.0^{8,10}]DODECANE**
CAS RN   105662-04-2

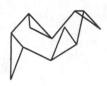

pentacyclo[9.1.0.0^{2,4}.0^{5,7}.0^{8,10}] dodecane 1α, 2β, 4β, 5α, 7α, 8β, 10β, 11α

ΔH_f^o = 61.4 kcal/mol

SPIRO[CYCLOBUTANE-1,3'-TETRACYCLO[3.3.1.0^{2,4}.0^{6,8}]NONANE]
CAS RN   170880-38-3

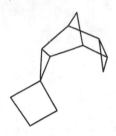

ΔH_f^o = 108.8 kcal/mol

SPIRO[CYCLOPENTANE-1,8'-TETRACYCLO[4.2.0.0^{1,7}.0^{2,7}]OCTANE]
CAS RN   120546-16-9

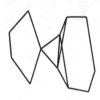

ΔH_f^o = 90.1 kcal/mol

SPIRO[CYCLOPROPANE-1,10'-TETRACYCLO[4.3.1.0^{2,4}.0^{7,9}]DECANE]
CAS RN   147085-41-4
              148021-27-6*

spiro[tetracyclo[4.3.1.0^{2,4}.0^{7,9}]decane-10,1'-cyclopropane] 1α, 2β, 4β, 6α, 7β, 9β

ΔH_f^o = 122.2 kcal/mol

SPIRO[CYCLOPROPANE-1,10'-TETRACYCLO[4.3.1.0^{2,9}.0^{3,5}]DECANE]
CAS RN   68152-41-0

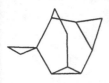

ΔH_f^o = 72.9 kcal/mol

***exo, exo*-SPIRO[TETRACYCLO[4.3.1.0^{2,5}.0^{7,9}]DECANE-3,1'-CYCLOPROPANE]**
CAS RN   126450-67-7

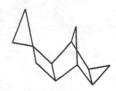

spiro[cyclopropane-1,3'-tetracyclo[4.3.1.0^{2,5}.0^{7,9}]decane] 1'α, 2'β, 5'β, 6'α, 7'β, 9'β

ΔH_f^o = 63.1 kcal/mol

TETRASPIRO[2.0.2.0.2.0.2.0]DODECANE
CAS RN   24375-17-5

[4]ROTANE
[4.3]ROTANE
TETRACYCLOPROPYLIDENE

ΔH_f^o = 83.4 kcal/mol

D. HEXACYCLODODECANES - $C_{12}H_{14}$

HEXACYCLO[5.4.1.0^{2,6}.0^{3,10}.0^{4,8}.0^{9,12}]DODECANE
CAS RN   704-02-9

BIRDCAGE HYDROCARBON
BISHOMOPENTAPRISMANE
1,3-BISHOMOPENTAPRISMANE
1,7-METHANOHOMOPENTAPRISMANE
1,5,2,4-ethanediylidenecyclopenta[cd]pentalene, decahydro

ΔH_f^o = 29.0 kcal/mol

HEXACYCLO[5.5.0.0^{1,8}.0^{2,4}.0^{3,5}.0^{7,9}]DODECANE
CAS RN   70528-31-3

1,2,3:4a,5,8a-dimethenonaphthalene, octahydro

ΔH_f^o = 125.1 kcal/mol

HEXACYCLO[5.5.0.0^{2,4}.0^{3,11}.0^{5,10}.0^{8,12}]DODECANE
CAS RN   76374-14-6

1,4:2,3-dimethanocyclobuta[a]cyclopropa[cd]pentalene, decahydro

ΔH_f^o = 46.9 kcal/mol

HEXACYCLO[5.5.0.0^{2,5}.0^{3,11}.0^{4,9}.0^{8,12}]DODECANE
CAS RN   72183-46-1

2,1,3-ethanylylidenedicyclobuta[cd,gh]pentalene, decahydro

ΔH_f^o = 43.6 kcal/mol

HEXACYCLO[6.4.0.0^{2,7}.0^{3,6}.0^{4,11}.0^{5,10}]DODECANE
CAS RN   106880-88-0

SECOHEXAPRISMANE
SECO[6]PRISMANE

ΔH_f^o = 73.8 kcal/mol

HEXACYCLO[6.4.0.0^{2,7}.0^{3,6}.0^{4,12}.0^{5,9}]DODECANE
CAS RN   62415-16-1

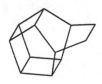

1,1'-BISHOMOPENTAPRISMANE
1,2,3,6-ethanediylidenecyclobut[cd]indene, decahydro

ΔH_f^o = 61.2 kcal/mol

HEXACYCLO[6.4.0.0^{2,7}.0^{3,12}.0^{4,6}.0^{9,11}]DODECANE
CAS RN   65071-73-0
                  65375-86-2*

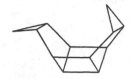

BISHOMOTETRAHYDROHYPOSTROPHENE
hexacyclo[6.4.0.0^{2,7}.0^{3,12}.0^{4,6}.0^{9,11}]dodecane 1α, 2α, 3α, 4α, 6α, 7α, 8α, 9β, 11β, 12α

ΔH_f^o = 98.0 kcal/mol

HEXACYCLO[6.4.0.0^{2,10}.0^{3,6}.0^{4,9}.0^{5,7}]DODECANE
CAS RN   99755-74-5

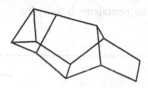

1,2,5-methano-1H-cyclopropa[3,4]cyclobut[1,2-a]pentalene, decahydro

ΔH_f^o = 60.7 kcal/mol

SPIRO[PENTACYCLO[4.4.0.0^{2,4}.0^{3,7}.0^{8,10}]DECANE-1,1'-CYCLOPROPANE]
CAS RN   167904-63-4

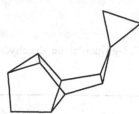

spiro[cyclopropane-1,1'-[2,3,5]metheno[1H]cyclopropa[a]pentalene], octahydro 1'α, 1'β,
2'α, 3'α, 4'aβ, 5'α, 5'aα

ΔH_f^o = 103.8 kcal/mol

E. HEPTACYCLODODECANES - $C_{12}H_{12}$

HEPTACYCLO[5.4.1.0^{1,10}.0^{2,6}.0^{3,10}.0^{4,8}.0^{9,12}]DODECANE
CAS RN   114095-92-0

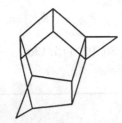

1,7-METHANOHOMOPENTAPRISMANE
5H-1,4b,2,4-ethanediylidene-1H-cyclopropa[1,4]cyclobuta[1,2,3'-cd]pentalene, hexahydro

ΔH_f^o = 105.7 kcal/mol

HEPTACYCLO[6.4.0.0^{2,4}.0^{3,7}.0^{5,12}.0^{6,10}.0^{9,11}]DODECANE
CAS RN   108639-57-2

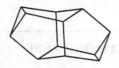

DITRIAXANE
p-[3^2.5^6]OCTAHEDRANE
1,2,3-metheno-1H-cyclopropa[cd]cyclopropa[4,5]cyclopenta[1,2,3-gh]pentalene, decahydro

ΔH_f^o = 72.1 kcal/mol

9
POLYCYCLOTRIDECANES
A. TRICYCLOTRIDECANES - C₁₃H₂₂

DISPIRO[5.0.5.1]TRIDECANE
CAS RN   174-34-5

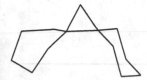

$\Delta H_f^o = -29.9$ kcal/mol

SPIRO[BICYCLO[2.2.1]HEPTANE-2,1'-CYCLOHEPTANE]
CAS RN   89944-09-2

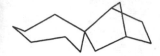

$\Delta H_f^o = -27.7$ kcal/mol

SPIRO[*cis*-BICYCLO[4.3.0]NONANE-1,1'-CYCLOPENTANE]
CAS RN   119140-78-2

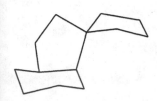

cis-spiro[cyclopentane-1,1'-[1H]indene, octahydro
$\Delta H_f^o = -38.0$ kcal/mol

SPIRO[*trans*-BICYCLO[4.3.0]NONANE-1,1'-CYCLOPENTANE]
CAS RN   119140-79-3

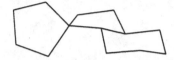

trans-spiro[cyclopentane-1,1'-[1H]indene], octahydro

ΔH_f^o = - 38.8 kcal/mol

TRICYCLO[4.4.3.0^{1,6}]TRIDECANE
CAS RN   3781-64-4

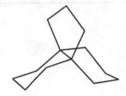

[4.4.3]PROPELLANE
4a,8a-propanonaphthalene, octahydro

ΔH_f^o = - 40.7 kcal/mol

TRICYCLO[6.5.0.0^{1,5}]TRIDECANE
CAS RN   119972-36-0

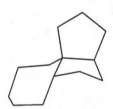

cyclopent[c]azulene, dodecahydro 3aα, 5aβ

ΔH_f^o = - 36.0 kcal/mol

TRICYCLO[6.5.0.0^{1,5}]TRIDECANE
CAS RN   120052-85-9

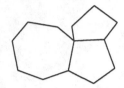

cyclopent[c]azulene, dodecahydro 3aα, 5aα

ΔH_f^o = - 35.5 kcal/mol

TRICYCLO[7.3.1.0^{1,6}]TRIDECANE
CAS RN   7156-77-6

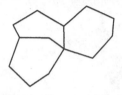

2H-4a,8-methanobenzocyclooctene, decahydro

ΔH_f^o = - 42.1 kcal/mol

TRICYCLO[7.3.1.0^{2,7}]TRIDECANE
CAS RN   80627-73-4
                80657-74-5*

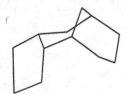

5,9-methanobenzocyclooctene, dodecahydro 4aα, 5α, 9α, 10aα

ΔH_f^o = - 36.7 kcal/mol

TRICYCLO[7.3.1.0^{2,7}]TRIDECANE
CAS RN   80627-71-2
               80657-74-5*

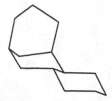

5,9-methanobenzocyclooctene, dodecahydro 4aα, 5α, 9α, 10aβ

ΔH_f^o = - 43.6 kcal/mol

TRICYCLO[7.3.1.0^{2,7}]TRIDECANE
CAS RN   80657-72-3
               80657-74-5*

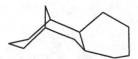

5,9-methanobenzocyclooctene, dodecahydro 4aα, 5β, 9β, 10aα

ΔH_f^o = - 41.1 kcal/mol

TRICYCLO[7.3.1.0^{2,7}]TRIDECANE
CAS RN   80627-88-9
               80657-74-5*

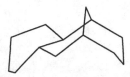

5,9-methanobenzocyclooctene, dodecahydro 4aα, 5β, 9β, 10aβ

ΔH_f^o = - 41.9 kcal/mol

---

cis,cis,cis-TRICYCLO[7.3.1.0⁵,¹³]TRIDECANE

cis,cis,cis-TRICYCLO[7.3.1.0^{5,13}]TRIDECANE
CAS RN   91465-60-0
            2935-07-1*

cis,cis,cis-PERHYDROPHENALENE
1H-phenalene, dodecahydro 3aα, 6aα, 9aα, 9bα

ΔH_f^o = - 37.8 kcal/mol

---

cis,cis,trans-TRICYCLO[7.3.1.0^{5,13}]TRIDECANE
CAS RN   86118-18-5
            2935-07-1*

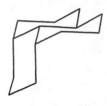

cis,cis,trans-PERHYDROPHENALENE
1H-phenalene, dodecahydro 3aα, 6aα, 9aβ, 9bα

ΔH_f^o = - 48.7 kcal/mol

---

cis,trans,trans-TRICYCLO[7.3.1.0^{5,13}]TRIDECANE
CAS RN   91465-59-7
            2935-07-1*

cis,trans,trans-PERHYDROPHENALENE
1H-phenalene, dodecahydro 3aα, 6aα, 9aβ, 9bβ

ΔH_f^o = - 42.3 kcal/mol

***trans,trans,trans*-TRICYCLO[7.3.1.0^{5,13}]TRIDECANE**
CAS RN   40250-64-4
            2935-07-1*

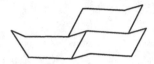

trans,trans,trans-PERHYDROPHENALENE
1H-phenalene, dodecahydro 3aα, 6aα, 9aα, 9bβ

ΔH_f^o = - 52.4 kcal/mol

***cis-cisoid-cis*-TRICYCLO[7.4.0.0^{2,6}]TRIDECANE**
CAS RN   77697-32-6
            30146-18-0*

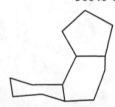

cis-syn-cis-1,2-CYCLOPENTADECALIN
1H-benz[e]indene, dodecahydro 3aα, 5aα, 9aα, 9bα

ΔH_f^o = - 39.1 kcal/mol

***cis-cisoid-trans*-TRICYCLO[7.4.0.0^{2,6}]TRIDECANE**
CAS RN   79926-84-4
            30146-18-0*

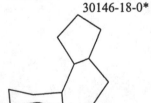

cis-syn-trans-1,2-PENTADECALIN
1H-benz[e]indene, dodecahydro 3aα, 5aβ, 9aβ, 9bβ

ΔH_f^o = - 42.1 kcal/mol

---

cis-transoid-cis-TRICYCLO[7.4.0.0^{2,6}]TRIDECANE

CAS RN   77697-34-8
            30146-18-0*

cis-anti-cis-1,2-CYCLOPENTADECALIN

1H-benz[e]indene, dodecahydro 3aα, 5aβ, 9aβ, 9bα

ΔH_f^o = - 42.3 kcal/mol

---

cis-transoid-trans-TRICYCLO[7.4.0.0^{2,6}]TRIDECANE

CAS RN   79926-83-3
            30146-18-0*

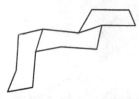

cis-anti-trans-1,2-CYCLOPENTADECALIN

1H-benz[e]indene, dodecahydro 3aα, 5aα, 9aα, 9bβ

ΔH_f^o = - 43.0 kcal/mol

---

trans-cisoid-cis-TRICYCLO[7.4.0.0^{2,6}]TRIDECANE

CAS RN   77697-36-0
            30146-18-0*

trans-syn-cis-1,2-CYCLOPENTADECALIN

1H-benz[e]indene, dodecahydro 3aα, 5aβ, 9aα, 9bα

ΔH_f^o = - 44.3 kcal/mol

***trans-cisoid-trans*-TRICYCLO[7.4.0.0²,⁶]TRIDECANE**
CAS RN   77697-35-9
                30146-18-0*

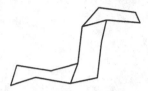

trans-syn-trans-1,2-CYCLOPENTADECALIN
1H-benz[e]indene, dodecahydro 3aα, 5aα, 9aβ, 9bβ

ΔH_f^o = - 41.0 kcal/mol

***trans-transoid-cis*-TRICYCLO[7.4.0.0²,⁶]TRIDECANE**
CAS RN   77697-33-7
                30146-18-0*

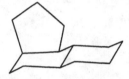

trans-anti-cis-1,2-CYCLOPENTADECALIN
1H-benz[e]indene, dodecahydro 3aα, 5aα, 9aβ, 9bα

ΔH_f^o = - 43.8 kcal/mol

***cis-cisoid-cis*-TRICYCLO[7.4.0.0²,⁷]TRIDECANE**
CAS RN   38106-06-8
                5744-03-6*

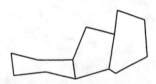

cis-syn-cis-PERHYDROFLUORENE
1H-fluorene, dodecahydro 4aα, 4bα, 8aα, 9aα

ΔH_f^o = - 40.0 kcal/mol

***cis-cisoid-trans*-TRICYCLO[7.4.0.0^{2,7}]TRIDECANE**
CAS RN   38106-04-6 (±)
              66605-55-8
              5744-03-6*

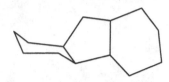

cis-syn-trans-PERHYDROFLUORENE
1H-fluorene, dodecahydro 4aα, 4bα, 8aα, 9aβ

ΔH_f^o = - 41.3 kcal/mol

***cis-transoid-cis*-TRICYCLO[7.4.0.0^{2,7}]TRIDECANE**
CAS RN   38106-05-7 (±)
              55619-69-7
              5744-03-6*

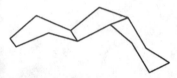

cis-anti-cis-PERHYDROFLUORENE
1H-fluorene, dodecahydro 4aα, 4bβ, 8aβ, 9aα

ΔH_f^o = - 41.3 kcal/mol

***trans-cisoid-trans*-TRICYCLO[7.4.0.0^{2,7}]TRIDECANE**
CAS RN   38106-03-5

trans-syn-trans-PERHYDROFLUORENE
1H-fluorene, dodecahydro 4aα, 4bα, 8aβ, 9aβ

ΔH_f^o = - 38.7 kcal/mol

***trans-transoid-cis*-TRICYCLO[7.4.0.0^{2,7}]TRIDECANE**
CAS RN   38106-02-4 (±)
               55619-68-6
               5744-03-6*

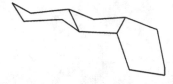

trans-anti-cis-PERHYDROFLUORENE
1H-fluorene, dodecahydro 4aα, 4bβ, 8aα, 9aα

ΔH_f^o = - 42.6 kcal/mol

***trans-transoid-trans*-TRICYCLO[7.4.0.0^{2,7}]TRIDECANE**
CAS RN   38106-01-3 (±)
               66605-54-7
               5744-03-6*

trans-anti-trans-PERHYDROFLUORENE
1H-fluorene, dodecahydro 4aα, 4bβ, 8aα, 9aβ

ΔH_f^o = - 43.0 kcal/mol

B. TETRACYCLOTRIDECANES - $C_{13}H_{20}$

DISPIRO[BICYCLO[2.2.1]HEPTANE-7,1'-CYCLOBUTANE-3',1''-CYCLOBUTANE]
CAS RN   63059-89-2

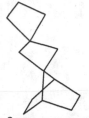

ΔH_f^o = 22.2 kcal/mol

7,7'-SPIROBIS[BICYCLO[4.1.0]HEPTANE]
CAS RN   181-43-1

7,7'-SPIROBI[BICYCLO[4.1.0]HEPTANE]
7,7'-SPIROBINORCARANE

ΔH_f^o = 21.4 kcal/mol

SPIRO[CYCLOBUTANE-1,2'-TRICYCLO[3.3.1.1^{3,7}]DECANE]
CAS RN   123489-44-1

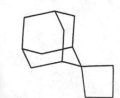

SPIRO[ADAMANTANE-2'-CYCLOBUTANE]
spiro[cyclobutane-1,6'-(2'H)-2,5-methano[1H]indene], hexahydro

ΔH_f^o = - 14.2 kcal/mol

TETRACYCLO[5.4.1.1^{2,6}.0^{4,9}]TRIDECANE
CAS RN 93832-14-5

HOMOICEANE
2,7:3,6-dimethano-1H-benzocycloheptene, decahydro

ΔH_f^o = - 18.9 kcal/mol

TETRACYCLO[5.5.1.0^{1,7}.0^{4,13}]TRIDECANE
CAS RN 80800-23-3

3H,6H-cyclobuta[1,3]cyclopenta[2,3]cyclopropa[1,2]cycloheptene, octahydro

ΔH_f^o = - 4.8 kcal/mol

***all-cis*-TETRACYCLO[5.5.1.0^{4,13}.0^{10,13}]TRIDECANE**
CAS RN 106566-70-5
67490-05-5*

TETRAQUINACANE
TETRAFUSOTETRAQUINANE
STAURANE
all-cis-[5.5.5.5]FENESTRANE
cis-1-transoid-1,4-cis-4-transoid-4,7-cis-7-transoid-7,10-cis-10-transoid-10,1-tetracyclo-
[5.5.1.0^{4,13}.0^{10,13}]tridecane
pentaleno[1,6-cd]pentalene, dodecahydro 2aα, 4aβ, 6aα, 8aβ

ΔH_f^o = - 16.0 kcal/mol

TETRACYCLO[6.2.2.1³,⁶.0²,⁷]TRIDECANE
CAS RN   81012-53-5

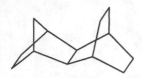

1,4-ethano-5,8-methanonaphthalene, decahydro

ΔH_f^o = - 15.1 kcal/mol

TETRACYCLO[6.3.1.1³,⁶.0²,⁷]TRIDECANE
CAS RN   58840-54-3

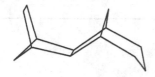

1,4:5,9-dimethano-1H-benzocycloheptene, decahydro 1α, 4α, 4aα, 5β, 9β, 9aα

ΔH_f^o = - 16.0 kcal/mol

TETRACYCLO[6.3.1.1³,⁶.0²,⁷]TRIDECANE
CAS RN   58865-58-0

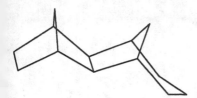

1,4:5,9-dimethano-1H-benzocycloheptene, decahydro 1α, 4α, 4aβ, 5α, 9α, 9aβ

ΔH_f^o = - 17.0 kcal/mol

TETRACYCLO[6.3.1.1^{6,10}.0^{1,5}]TRIDECANE

CAS RN   52353-12-5   123669-95-4 (S)

              123489-42-9 (±)   123618-32-6 (R)

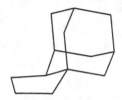

1,2-CYCLOPENTANOADAMANTANE
1,2-TRIMETHYLENEADAMANTANE
3a,7:5,9-dimethano-3aH-cyclopentacyclooctene, decahydro

ΔH_f^o = - 32.3 kcal/mol

TETRACYCLO[7.3.1.0^{2,5}.0^{5,10}]TRIDECANE

CAS RN   57387-00-5

3,6-methano-1H-cyclobuta[d]naphthalene, decahydro

ΔH_f^o = - 1.2 kcal/mol

TETRACYCLO[7.3.1.0^{2,7}.0^{6,11}]TRIDECANE

CAS RN   74475-01-7

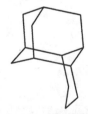

endo-2,4-TRIMETHYLENEADAMANTANE
1,5,3-[1,2,3]propanetriylnaphthalene, decahydro

ΔH_f^o = - 31.3 kcal/mol

TETRACYCLO[8.2.1.0²,⁹.0⁴,¹¹]TRIDECANE

CAS RN 100762-64-9 (±)
100693-75-2 (S)

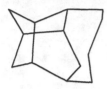

[4.1]TRIBLATTANE
1,5:2,4-dimethano-1H-cyclopentacyclooctene, decahydro 1α, 2α, 3aβ, 4α, 5α, 9aβ

ΔH_f^o = 3.0 kcal/mol

TETRACYCLO[8.2.1.0⁴,¹².0⁷,¹¹]TRIDECANE

CAS RN 112843-00-2

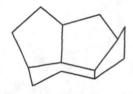

1H-cyclopent[jkl]-*as*-indacene, dodecahydro

ΔH_f^o = - 24.5 kcal/mol

C. PENTACYCLOTRIDECANES - $C_{13}H_{18}$

endo,exo,syn-PENTACYCLO[4.4.3.0^{2,5}.0^{7,10}.0^{11,13}]TRIDECANE

CAS RN   75813-58-0

              76035-39-7*

ΔH_f^o = 59.6 kcal/mol

PENTACYCLO[5.5.1.0^{1,7}.0^{3,5}.0^{9,11}]TRIDECANE

CAS RN   27714-85-8

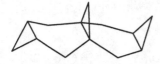

1H,3H-2a,5a-methanodicyclopropa[bg]naphthalene, octahydro

ΔH_f^o = 44.0 kcal/mol

PENTACYCLO[6.3.1.1^{3,6}.0^{2,7}.0^{9,11}]TRIDECANE

CAS RN   61140-68-9

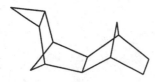

ΔH_f^o = 27.8 kcal/mol

194

PENTACYCLO[6.5.0.0^{2,5}.0^{3,10}.0^{4,9}]TRIDECANE
CAS RN   94319-69-4

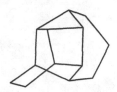

[3.2.0]TRIBLATTANE
1,9,2,5-ethanediylidene-1H-benzocycloheptene, decahydro

ΔH_f^o = 24.9 kcal/mol

PENTACYCLO[6.5.0.0^{2,6}.0^{3,10}.0^{5,9}]TRIDECANE
CAS RN   88400-60-6 (-)

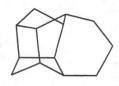

[3.1.1]TRIBLATTANE
1,3,5-methenocyclopent[cd]azulene, dodecahydro

ΔH_f^o = 9.0 kcal/mol

PENTACYCLO[6.5.0.0^{2,11}.0^{3,10}.0^{9,12}]TRIDECANE
CAS RN   94319-67-2

[4.1.0]TRIBLATTANE
1,9,2,4-ethanediylidene-1H-cyclopentacyclooctene, decahydro 1α, 2α, 3aβ, 4α, 9α, 9aβ

ΔH_f^o = 43.5 kcal/mol

PENTACYCLO[6.5.0.0^{4,13}.0^{5,9}.0^{10,12}]TRIDECANE
CAS RN   81012-55-7

3,2,6-[1]propanyl[3]ylidenecycloprop[e]indene, decahydro

ΔH_f^o = 22.6 kcal/mol

PENTACYCLO[7.3.1.0^{2,4}.0^{3,8}.0^{5,11}]TRIDECANE
CAS RN   93832-12-3

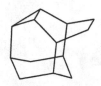

3,13-DEHYDROICEANE
CYCLOHOMOICEANE
3,4-ethano-2,5-methanocyclopropa[de]naphthalene, decahydro

ΔH_f^o = 6.9 kcal/mol

PENTACYCLO[7.3.1.0^{2,8}.0^{3,5}.0^{10,12}]TRIDECANE
CAS RN   151291-87-1

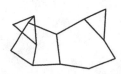

4,6-methanodicyclopropa[1,g]naphthalene, dodecahydro

ΔH_f^o = 58.4 kcal/mol

PENTACYCLO[7.4.0.0^{2,6}.0^{3,11}.0^{5,10}]TRIDECANE

CAS RN   153744-27-5   116347-49-0 (1R)
           70224-69-0 (±)   77122-03-3 (1S)

[2.2.1]TRIBLATTANE
METHANODITWISTANE
1,3,6-methenoacenaphthylene, dodecahydro

ΔH_f^o = - 3.0 kcal/mol

SPIRO[CYCLOHEXANE-1,8'-TETRACYCLO[4.2.0.0^{1,7}.0^{2,7}]OCTANE]

CAS RN   120546-14-7

ΔH_f^o = 78.7 kcal/mol

TETRASPIRO[2.0.2.0.2.0.2.1]TRIDECANE

CAS RN   24971-87-7

ΔH_f^o = 45.0 kcal/mol

TRISPIRO[CYCLOPROPANE-1,2'-BICYCLO[2.2.1]HEPTANE-5',1''-CYCLO-PROPANE-7',1'''-CYCLOPROPANE]
CAS RN   127429-97-4

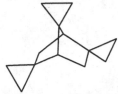

trispiro[bicyclo[2.2.1]heptane-2,1':5,1'':7,1'''-triscyclopropane]

ΔH_f^o = 62.8 kcal/mol

TRISPIRO[CYCLOPROPANE-1,2'-BICYCLO[2.2.1]HEPTANE-6',1''-CYCLOPRO-PANE-7',1'''-CYCLOPROPANE]
CAS RN   127429-98-5

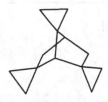

trispiro[bicyclo[2.2.1]heptane-2,1':6,1'':7,1'''-triscyclopropane]

ΔH_f^o = 67.9 kcal/mol

D. HEXACYCLOTRIDECANES - $C_{13}H_{16}$

HEXACYCLO[4.4.3.0^{2,10}.0^{3,5}.0^{7,9}.0^{11,13}]TRIDECANE
CAS RN   31859-55-9
            59246-25-2 (+)
            59246-26-3 (±)

TRISHOMOBULLVALENE

ΔH_f^o = 90.0 kcal/mol

HEXACYCLO[5.5.1.0^{2,6}.0^{3,10}.0^{4,8}.0^{9,13}]TRIDECANE
CAS RN   71871-52-8

HOMOBIRDCAGE HYDROCARBON
1,5,2,4-ethanediylidene-1H-cyclopent[cd]indene, decahydro

ΔH_f^o = 19.4 kcal/mol

HEXACYCLO[5.5.1.0^{2,6}.0^{3,11}.0^{4,9}.0^{10,13}]TRIDECANE
CAS RN   24856-73-3

1,5,2,4-[1,2]propanediyl[3]ylidenecyclopenta[cd]pentalene, decahydro
ΔH_f^o = 5.5 kcal/mol

HEXACYCLO[6.4.1.0^{2,7}.0^{4,11}.0^{5,9}.0^{10,13}]TRIDECANE
CAS RN   127530-76-1

TRISHOMOPENTAPRISMANE

1,6:2,5:3,4-trimethanocyclobuta[1,2:3,4]dicyclopentene, decahydro

ΔH_f^o = 24.4 kcal/mol

HEXACYCLO[7.3.1.0^{2,4}.0^{3,7}.0^{5,11}.0^{6,8}]TRIDECANE
CAS RN   93832-13

2,3-methanodicycloprop[bc,jk]acenaphthylene, dodecahydro

ΔH_f^o = 63.0 kcal/mol

---

endo,exo,exo-HEXACYCLO[7.3.1.0^{2,4}.0^{5,13}.0^{6,8}.0^{10,12}]TRIDECANE
CAS RN   55820-80-9

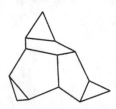

endo,exo,exo-TRISHOMOTRIQUINACENE

1H-dicyclopropa[a,f]cyclopropa[4,5]cyclopenta[1,2,3-cd]pentalene, dodecahydro 1aα, 1bβ, 1cα, 2aα, 2bβ, 2cβ, 3aβ, 3bβ, 3cα, 3dβ

ΔH_f^o = 90.4 kcal/mol

***exo,exo,exo*-HEXACYCLO[7.3.1.0^{2,4}.0^{5,13}.0^{6,8}.0^{10,12}]TRIDECANE**
CAS RN   55756-71-3

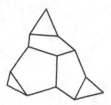

all-exo-TRISHOMOTRIQUINACENE
1H-dicyclopropa[a,f]cyclopropa[4,5]cyclopenta[1,2,3-cd]pentalene, dodecahydro 1aα, 1bβ, 1cα, 2aα, 2bβ, 2cα, 3aα, 3bβ, 3cα, 3dβ

ΔH_f^o = 89.6 kcal/mol

***cis*-HEXACYCLO[10.1.0.0^{1,3}.0^{3,5}.0^{5,7}.0^{7,9}]TRIDECANE**
CAS RN   159618-78-7*

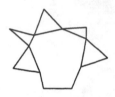

ΔH_f^o = 136.7 kcal/mol

***trans*-HEXACYCLO[10.1.0.0^{1,3}.0^{3,5}.0^{5,7}.0^{7,9}]TRIDECANE**
CAS RN   159618-78-7*

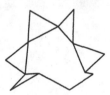

ΔH_f^o = 136.5 kcal/mol

PENTASPIRO[2.0.0.0.0.2.1.1.1.1]TRIDECANE
CAS RN   129872-33-9 (R)
                129940-07-4 (S)
                130464-17-4*

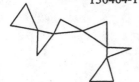

ΔH_f^o = 164.0 kcal/mol

PENTASPIRO[2.0.0.0.2.0.2.0.1.1]TRIDECANE
CAS RN   136504-51-3

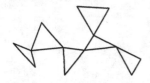

[6]TRIANGULANE

ΔH_f^o = 166.8 kcal/mol

---

cis-**PENTASPIRO[2.0.0.0.2.1.0.2.0.1]TRIDECANE**
CAS RN   136028-57-4
                136677-81-1*

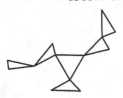

cis-[6]TRIANGULANE

ΔH_f^o = 164.1 kcal/mol

---

trans-PENTASPIRO[2.0.0.0.2.1.0.2.0.1]TRIDECANE
CAS RN   136028-46-1
             136677-81-1*

trans-[6]TRIANGULANE

ΔH_f^o = 170.1 kcal/mol

PENTASPIRO[2.0.0.2.0.2.0.0.2.0]TRIDECANE
CAS RN   137943-72-7

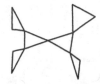

ΔH_f^o = 162.4 kcal/mol

SPIRO[CYCLOPROPANE-1,11'-PENTACYCLO[4.4.1.0^{2,10}.0^{3,5}.0^{7,9}]UNDECANE]
CAS RN   67654-16-4
             68107-93-7*

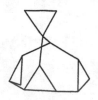

endo,exo-BISHOMOBARBARALANE-9-SPIROCYCLOPROPANE

ΔH_f^o = 113.7 kcal/mol

SPIRO[CYCLOPROPANE-1,11'-PENTACYCLO[4.4.1.0^{2,10}.0^{3,5}.0^{7,9}]UNDECANE]
CAS RN   67672-87-1
               68107-93-7*

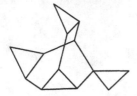

exo,exo-BISHOMOBARBARALANE-9-SPIROCYCLOPROPANE

ΔH_f^o = 112.2 kcal/mol

SPIRO[CYCLOPROPANE-1,4'-PENTACYCLO[6.3.0.0^{2,6}.0^{3,10}.0^{5,9}]UNDECANE]
CAS RN   156033-23-7 (±)

spiro[cyclopropane-1,2'(1'H)-[1,3,5]methenocyclopenta[cd]pentalene, octahydro

ΔH_f^o = 36.9 kcal/mol

**TRISPIRO[CYCLOPROPANE-1,3'-TRICYCLO[2.2.1.0^{2,6}]HEPTANE]-5',1''-
CYCLOPROPANE-(7',1''')-CYCLOPROPANE**
CAS RN   127429-96-3

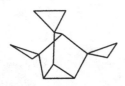

TRISPIROCYCLOPROPANENORTRICYCLANE
TRIS(SPIROCYCLOPROPYL)NORTRICYCLANE
trispiro[tricyclo[2.2.1.0^{2,6}]heptane-3,1':5,1'':7,1'''-triscyclopropane]

ΔH_f^o = 97.3 kcal/mol

E. HEPTACYCLOTRIDECANES - $C_{13}H_{14}$

HEPTACYCLO[5.5.1.0^{1,7}.0^{2,4}.0^{3,5}.0^{8,10}.0^{9,11}]TRIDECANE
CAS RN   73320-95-3

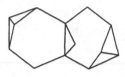

4a,8a-methano-1,2,3:5,6,7-dimethenonaphthalene, octahydro

ΔH_f^o = 141.8 kcal/mol

HEPTACYCLO[5.5.1.0^{1,7}.0^{2,4}.0^{3,5}.0^{9,11}.0^{10,12}]TRIDECANE
CAS RN   73320-94-2

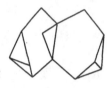

4a,8a-methano-1,2,3:6,7,8-dimethenonaphthalene, octahydro

ΔH_f^o = 142.4 kcal/mol

10
POLYCYCLOTETRADECANES
A. TRICYCLOTETRADECANES - $C_{14}H_{24}$

DISPIRO[5.1.5.1]TETRADECANE
CAS RN   184-97-4

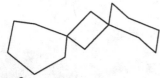

ΔH_f^o = - 32.7 kcal/mol

SPIRO[BICYCLO[4.2.1]NONANE-9,1'-CYCLOHEXANE]
CAS RN   137124-10-8

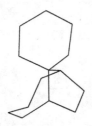

ΔH_f^o = - 37.1 kcal/mol

TRICYCLO[4.4.4.0^{1,6}]TETRADECANE
CAS RN   13755-04-9

[4.4.4]PROPELLANE
4a,8a-butanonaphthalene, octahydro

ΔH_f^o = - 48.2 kcal/mol

TRICYCLO[6.3.3.0^{1,8}]TETRADECANE
CAS RN   67140-86-7

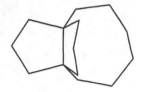

[6.3.3]PROPELLANE
3a,9a-propano-1H-cyclopentacyclooctene, octahydro

ΔH_f^o = - 31.5 kcal/mol

---

cis-cisoid-cis-TRICYCLO[6.6.0.0^{2,7}]TETRADECANE
CAS RN   53749-77-2

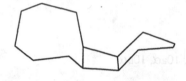

benzo[3,4]cyclobuta[1,2]cyclooctene, tetradecahydro 4aα, 4bα, 10aα, 10bα

ΔH_f^o = - 20.6 kcal/mol

---

cis-transoid-trans-TRICYCLO[6.6.0.0^{2,7}]TETRADECANE
CAS RN   76024-46-9

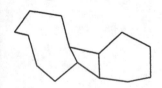

benzo[3,4]cyclobuta[1,2]cyclooctene, tetradecahydro 4aα, 4βα, 10aβ, 10bα

ΔH_f^o = - 23.4 kcal/mol

TRICYCLO[7.4.1.0^{4,14}]TETRADECANE
CAS RN   830-56-8

cyclopent[ef]heptalene, tetradecahydro

ΔH_f^o = - 40.0 kcal/mol

cis-cisoid-cis-TRICYCLO[7.5.0.0^{2,8}]TETRADECANE
CAS RN   17385-38-5

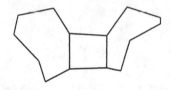

cyclobuta[1,2:3,4]dicycloheptene, tetradecahydro 5aα, 5bα, 10aα, 10bα

ΔH_f^o = - 24.1 kcal/mol

cis-transoid-cis-TRICYCLO[7.5.0.0^{2,8}]TETRADECANE
CAS RN   34737-51-4

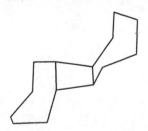

cyclobuta[1,2:3,4]dicycloheptene, tetradecahydro 5aα, 5bβ, 10aβ, 10bα

ΔH_f^o = - 26.5 kcal/mol

trans-cisoid-trans-TRICYCLO[7.5.0.0^{2,8}]TETRADECANE
CAS RN 83350-02-1

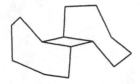

cyclobuta[1,2:3,4]dicycloheptene, tetradecahydro 5aα, 5bα, 10aβ, 10bβ

ΔH_f^o = - 14.4 kcal/mol

trans-transoid-trans-TRICYCLO[7.5.0.0^{2,8}]TETRADECANE
CAS RN 51607-15-9

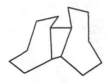

cyclobuta[1,2:3,4]dicycloheptene, tetradecahydro 5aα, 5bβ, 10aα, 10bβ

ΔH_f^o = - 24.4 kcal/mol

trans,cis-TRICYCLO[8.4.0.0^{1,6}]TETRADECANE
CAS RN 62672-97-3

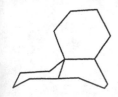

trans,cis-1H-benzo[d]naphthalene,dodecahydro

ΔH_f^o = - 50.5 kcal/mol

cis-cisoid-cis-TRICYCLO[8.4.0.0^{2,7}]TETRADECANE

CAS RN   26634-41-3
                 5743-97-5*

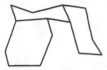

cis-syn-cis-PERHYDROPHENANTHRENE

phenanthrene, tetradecahydro 4aα, 4bα, 8aα, 10aα

ΔH_f^o = - 48.2 kcal/mol

---

cis-tranoid-cis-TRICYCLO[8.4.0.0^{2,7}]TETRADECANE

CAS RN   27389-74-8
                 5743-97-5*

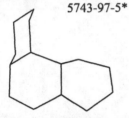

cis-anti-cis-PERHYDROPHENANTHRENE

phenanthrene, tetradecahydro 4aα, 4bβ, 8aβ, 10aα

ΔH_f^o = - 51.1 kcal/mol

---

trans-cisoid-cis-TRICYCLO[8.4.0.0^{2,7}]TETRADECANE

CAS RN   27425-35-0
                 5743-97-5*
                 38113-30-3 (±)

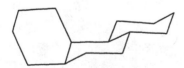

trans-syn-cis-PERHYDROPHENANTHRENE

phenanthrene, tetradecahydro 4aα, 4bα, 8aα, 10aβ

ΔH_f^o = - 54.1 kcal/mol

***trans-cisoid-trans*-TRICYCLO[8.4.0.0²·⁷]TETRADECANE**
CAS RN   27389-76-0
              5743-97-5*

***trans-syn-trans*-PERHYDROPHENANTHRENE**
phenanthrene, tetradecahydro 4aα, 4bα, 8aβ, 10aβ

ΔH_f^o = - 52.5 kcal/mol

***trans-transoid-cis*-TRICYCLO[8.4.0.0²·⁷]TETRADECANE**
CAS RN   27389-73-7   38113-29-0 (±)
              5743-97-5*   2108-89-6

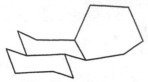

***trans-anti-cis*-PERHYDROPHENANTHRENE**
phenanthrene, tetradecahydro 4aα, 4bβ, 8aα, 10aα

ΔH_f^o = - 54.0 kcal/mol

***trans-transoid-trans*-TRICYCLO[8.4.0.0²·⁷]TETRADECANE**
CAS RN   38113-28-9 (±)
              5743-97-5*
              2108-89-6*

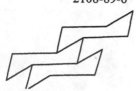

***trans-anti-trans*-PERHYDROPHENANTHRENE**
phenanthrene, tetradecahydro 4aα, 4bβ, 8aα, 10aβ

ΔH_f^o = - 56.6 kcal/mol

***cis-cisoid-cis*-TRICYCLO[8.4.0.0^{3,8}]TETRADECANE**
CAS RN   19128-78-0
          6596-35-6*

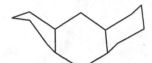

cis-syn-cis-PERHYDROANTHRACENE
anthracene, tetradecahydro 4aα, 8aα, 9aα, 10aα

ΔH_f^o = - 50.4 kcal/mol

***cis-cisoid-trans*-TRICYCLO[8.4.0.0^{3,8}]TETRADECANE**
CAS RN   2109-05-9
          29863-90-9 (±)
          6596-35-6*

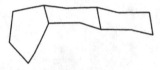

cis-syn-trans-PERHYDROANTHRACENE
anthracene, tetradecahydro 4aα, 8aα, 9aα, 10aβ

ΔH_f^o = - 55.4 kcal/mol

***cis-transoid-cis*-TRICYCLO[8.4.0.0^{3,8}]TETRADECANE**
CAS RN   29863-91-0 (±)
          6596-35-6*

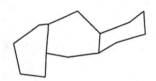

cis-anti-cis-PERHYDROANTHRACENE
anthracene, tetradecahydro 4aα, 8aβ, 9aα, 10aβ

ΔH_f^o = - 52.5 kcal/mol

---

trans-cisoid-trans-TRICYCLO[8.4.0.0^{3,8}]TETRADECANE
CAS RN   1755-19-7
                        6596-35-6*

trans-syn-trans-PERHYDROANTHRACENE
anthracene, tetradecahydro 4aα, 8aβ, 9aβ, 10aα

ΔH_f^o = - 58.1 kcal/mol

---

trans-transoid-trans-TRICYCLO[8.4.0.0^{3,8}]TETRADECANE
CAS RN   28071-99-0
                        30008-95-8 (±)
                        6596-35-6*

trans-anti-trans-PERHYDROANTHRACENE
anthracene, tetradecahydro 4aα, 8aα, 9aβ, 10aβ

ΔH_f^o = - 52.0 kcal/mol

TRICYCLO[9.3.0.0^{3,7}]TETRADECANE
CAS RN   69618-30-0*

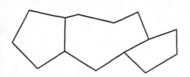

dicyclopenta[a,d]cyclooctene, tetradecahydro

ΔH_f^o = - 37.6 kcal/mol

---

cis,trans-TRICYCLO[11.1.0.0⁴,⁶]TETRADECANE

CAS RN   19651-17-3
                  3105-36-0*

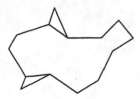

ΔH_f^o = - 2.1 kcal/mol

---

trans,trans-TRICYCLO[11.1.0.0⁴,⁶]TETRADECANE

CAS RN   19651-16-2
                  3105-36-0*

ΔH_f^o = - 4.3 kcal/mol

B. TETRACYCLOTETRADECANES - $C_{14}H_{22}$

anti-DISPIRO[BICYCLO[4.1.0]HEPTANE-7,1'-CYCLOPROPANE-2',1"-CYCLOHEXANE]
CAS RN   56453-42-0
                56760-99-7*

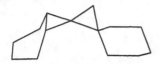

dispiro[bicyclo[4.1.0]heptane-7,1'-cyclopropane-2',1"-cyclohexane] 1α, 6α, 7α

ΔH_f^o = 16.8 kcal/mol

---

syn-DISPIRO[BICYCLO[4.1.0]HEPTANE-7,1'-CYCLOPROPANE-2',1"-CYCLOHEXANE]
CAS RN   56424-65-8
                56760-99-7*

dispiro[bicyclo[4.1.0]heptane-7,1'-cyclopropane-2',1"-cyclohexane] 1α, 6α, 7β

ΔH_f^o = 27.3 kcal/mol

DISPIRO[CYCLOHEXANE-1,2'-BICYCLO[1.1.0]BUTANE-4',1"-CYCLOHEXANE]
CAS RN   107924-94-7

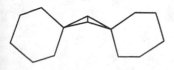

dispiro[cyclohexane-1,2'-bicyclo[1.1.0]butane-4',1"-cyclohexane]

ΔH_f^o = 29.4 kcal/mol

TETRACYCLO[5.5.2.0^{1,8}.0^{4,8}]TETRADECANE
CAS RN   119182-92-2

1,5a-ethano-5aH-cyclopent[c]indene, decahydro

ΔH_f^o = - 32.5 kcal/mol

TETRACYCLO[5.5.2.0^{4,13}.0^{10,14}]TETRADECANE
CAS RN   108195-85-3

PERHYDRODIACENAPHTHALENE
cyclopent[fg]acenaphthylene, tetradecahydro

ΔH_f^o = - 19.5 kcal/mol

TETRACYCLO[6.2.2.2^{3,6}.0^{1,6}]TETRADECANE
CAS RN   71826-18-1

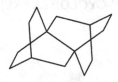

[2.2.2]^{2}GEMINANE
2,4a:6,8a-diethanonaphthalene, octahydro

ΔH_f^o = - 28.5 kcal/mol

exo,exo-TETRACYCLO[6.2.2.2^{3,6}.0^{2,7}]TETRADECANE
CAS RN   72107-00-7

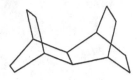

1,4:5,8-diethanonaphthalene, decahydro

ΔH_f^o = - 22.3 kcal/mol

TETRACYCLO[6.5.1.0^{2,7}.0^{9,13}]TETRADECANE
CAS RN   66807-96-3
            52353-11-4*

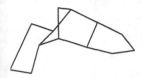

4,9-methano-1H-benz[f]indene, dodecahydro   3aα, 4β, 4aα, 8aα, 9β, 9aα

ΔH_f^o = - 1.04 kcal/mol

TETRACYCLO[6.5.1.0^{2,7}.0^{9,13}]TETRADECANE
CAS RN   66807-94-1
            52353-11-4*

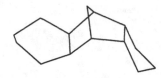

4,9-methano-1H-benz[f]indene, dodecahydro 3aα, 4α, 4aβ, 8aβ, 9α, 9aα

ΔH_f^o = - 22.2 kcal/mol

cis-cisoid-cis-TETRACYCLO[6.6.0.0^{1,11}.0^{3,7}]TETRADECANE
CAS RN   132883-19-3 (±)

1H-dicyclopenta[a,d]pentalene, dodecahydro   3aα, 5aβ, 5bβ, 8aβ

ΔH_f^o = - 28.3 kcal/mol

---

cis-transoid-cis-TETRACYCLO[6.6.0.0^{1,11}.0^{3,7}]TETRADECANE
CAS RN   132842-75-2 (±)

1H-dicyclopenta[a,d]pentalene, dodecahydro   3aα, 5aβ, 5bα, 8aα

ΔH_f^o = - 18.2 kcal/mol

---

all-cis-TETRACYCLO[6.6.0.0^{1,11}.0^{4,8}]TETRADECANE
CAS RN   102064-97-1*

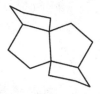

dicyclopenta[c,g]pentalene, dodecahydro

ΔH_f^o = - 31.6 kcal/mol

TETRACYCLO[7.3.1.1^{7,11}.0^{1,6}]TETRADECANE
CAS RN   41171-94-2

1,2-TETRAMETHYLENEADAMANTANE
CYCLOHEXANOADAMANTANE
2H-4a,8:6,10-dimethanobenzocyclooctene, decahydro

ΔH_f^o = - 43.8 kcal/mol

TETRACYCLO[8.2.1.1^{4,7}.0^{1,7}]TETRADECANE
CAS RN   71826-17-0

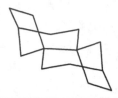

syn[3.2.1]^{2}GEMINANE
1H,6H-3,5a:8,10a-dimethanoheptalene, octahydro 3α, 5aα, 8α, 10aα

ΔH_f^o = - 28.9 kcal/mol

TETRACYCLO[9.2.1.0^{2,10}.0^{3,7}]TETRADECANE
CAS RN   33386-83-3

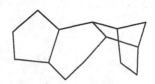

6,9-methano-1H-benz[e]indene, dodecahydro 3aα, 5aβ, 6β, 9β, 9aβ, 9bα

ΔH_f^o = - 22.6 kcal/mol

TETRACYCLO[9.2.1.0^{2,10}.0^{3,8}]TETRADECANE
CAS RN   100578-95-8

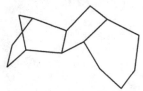

1,4-methano-1H-fluorene, dodecahydro 1α, 4α, 4aβ, 4bα, 8aα, 9aβ

ΔH_f^o = - 28.4 kcal/mol

TETRACYCLO[9.2.1.0^{2,10}.0^{3,8}]TETRADECANE
CAS RN   100680-28-2

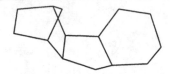

1,4-methano-1H-fluorene, dodecahydro 1α, 4α, 4aβ, 4bβ, 8aβ, 9αβ

ΔH_f^o = - 25.0 kcal/mol

TETRACYCLO[9.2.1.0^{2,10}.0^{4,12}]TETRADECANE
CAS RN   100762-65-0 (±)
               100693-76-3

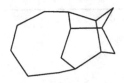

[5.1]TRIBLATTANE
1,5;2,4-dimethanocyclopentacyclononene, dodecahydro

ΔH_f^o = 3.0 kcal/mol

C. PENTACYCLOTETRADECANES - C₁₄H₂₀

DISPIRO[CYCLOPENTANE-1,3'-*trans*-TRICYCLO[3.1.0.0^{2,4}]HEXANE-6',1''-CYCLOPENTANE]
CAS RN   78578-93-5
                    79355-47-8

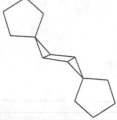

anti-anti-anti-tetraspiro[pentacylco[9.1.0.0^{2,4}.0^{5,7}.0^{8,10}]dodecane-3,1':6,1'':9,1'''-tetrakis-cyclopentane]
trispiro[tetracyclo[6.1.0.0^{2,4}.0^{5,7}]nonane-3,1':6,1'':9,1'''-triscyclopentane] 1'α, 2'β, 4'β, 5'α

ΔH_f^o = 19.2 kcal/mol

DISPIRO[CYCLOPROPANE-1,3'-TRICYCLO[5.2.1.0^{2,6}]DECANE-10',1''-CYCLOPROPANE]
CAS RN   105883-30-5

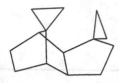

dispiro[cyclopropane-1,1'-[4,7]methano[1H]indene-8',1''-cyclopropane], octahydro
3'aα, 4'α, 7'α, 7'aα

ΔH_f^o = 31.2 kcal/mol

PENTACYCLO[5.5.1.1^{1,5}.0^{2,10}.0^{3,8}]TETRADECANE
CAS RN   60526-52-5

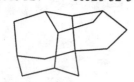

2H-3,5:3,7-dimethanocyclopent[a]indene, decahydro

ΔH_f^o = - 13.3 kcal/mol

PENTACYCLO[6.4.1.1^{2,7}.0^{1,8}.0^{2,7}]TETRADECANE
CAS RN   85739-37-3

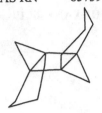

4a,8b:4b,8a-dimethanobiphenylene, octahydro 4aα, 4bβ, 8aβ, 8bα

ΔH_f^o = 29.2 kcal/mol

PENTACYCLO[6.6.0.0^{2,5}.0^{3,10}.0^{4,9}]TETRADECANE
CAS RN   94319-70-7 (1S)

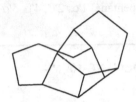

[4.2.0]TRIBLATTANE
1,10,2,5-ethanediylidenebenzocyclooctene, dodecahydro

ΔH_f^o = 29.2 kcal/mol

PENTACYCLO[6.6.0.0^{2,6}.0^{3,10}.0^{5,9}]TETRADECANE
CAS RN   88400-61-7 (-)

[4.1.1]TRIBLATTANE
1,3,5-metheno-1H-cycloocta[cd]pentalene, dodecahydro

ΔH_f^o = 14.0 kcal/mol

---

cis-PENTACYCLO[7.3.1.1^{3,7}.0^{1,9}.0^{3,7}]TETRADECANE
CAS RN   6555-86-8

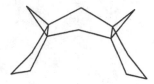

syn-dimethanoperhydro-s-hydrinacene

1H, 4H, 5H, 8H-3a, 8a:4a,7a-dimethano-s-indacene, tetrahydro 3aα, 4aα, 7aα, 8aα

ΔH_f^o = 16.3 kcal/mol

---

trans-PENTACYCLO[7.3.1.1^{3,7}.0^{1,9}.0^{3,7}]TETRADECANE
CAS RN   13221-77-7

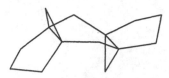

anti-dimethanoperhydro-s-hydrinacene

1H, 4H, 5H, 8H-3a, 8a:4a,7a-dimethano-s-indacene, tetrahydro 3aα, 4aβ, 7aβ, 8aα

ΔH_f^o = 15.6 kcal/mol

PENTACYCLO[7.3.1.1^{4,12}.0^{1,7}.0^{8,11}]TETRADECANE
CAS RN   42392-27-8

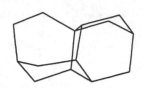

PROTODIAMANTANE

2,5:4,7-dimethano-1H-cyclopent[c]indene, decahydro

ΔH_f^o = - 16.7 kcal/mol

PENTACYCLO[7.5.0.0²,⁵.0³,¹¹.0⁴,¹⁰]TETRADECANE
CAS RN   94319-75-2 (1S)

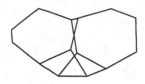

[3.3.0]TRIBLATTANE
1,2,6-metheno-1H-cyclobuta[ef]heptalene, dodecahydro

ΔH_f^o = 20.7 kcal/mol

PENTACYCLO[7.5.0.0²,⁸.0³,¹³.0⁵,¹⁰]TETRADECANE
CAS RN   79772-15-9

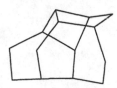

2,4-ethano-1,5-methano-1H-cyclopenta[3,4]cyclobuta[1,2]benzene, decahydro

ΔH_f^o = 11.2 kcal/mol

PENTACYCLO[7.5.0.0²,⁸.0⁴,¹².0⁵,¹¹]TETRADECANE
CAS RN   7018-56-6

ΔH_f^o = 32.2 kcal/mol

PENTACYCLO[7.5.0.0²,¹².0³,¹¹.0¹⁰,¹³]TETRADECANE

$PENTACYCLO[7.5.0.0^{2,12}.0^{3,11}.0^{10,13}]TETRADECANE$

CAS RN   94319-68-3 (1S)

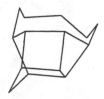

[5.1.0]TRIBLATTANE
1,10,2,4-ethanediylidenecyclopentacyclononene, dodecahydro

ΔH_f^o = 43.1 kcal/mol

---

cis-PENTACYCLO[8.2.1.1²,⁵.0³,⁷.0⁸,¹²]TETRADECANE

$cis\text{-}PENTACYCLO[8.2.1.1^{2,5}.0^{3,7}.0^{8,12}]TETRADECANE$

CAS RN   52021-60-0

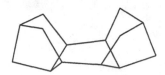

TETRAHYDROBINOR-S
2,4:6,8-dimethano-s-indacene, dodecahydro 2α, 3aβ, 4α, 4aβ, 6α, 7aβ, 8α, 8aβ

ΔH_f^o = - 10.9 kcal/mol

---

trans-PENTACYCLO[8.2.1.1²,⁵.0³,⁷.0⁸,¹²]TETRADECANE

$trans\text{-}PENTACYCLO[8.2.1.1^{2,5}.0^{3,7}.0^{8,12}]TETRADECANE$

CAS RN   51966-16-6

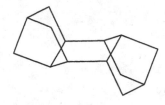

TETRAHYDROBINOR-A
2,4:6,8-dimethano-s-indacene, dodecahydro 2α, 3aβ, 4α, 4aβ, 6β, 7aβ, 8β, 8aβ

ΔH_f^o = 1.04 kcal/mol

D. HEXACYCLOTETRADECANES - $C_{14}H_{18}$

HEXACYCLO[6.5.1.0^{2,6}.0^{3,12}.0^{5,10}.0^{9,13}]TETRADECANE
CAS RN   61248-60-0

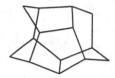

5,3,2,6-[1,2]propanediyl[3]ylidene-1H-cyclopent[cd]indene, dodecahydro

ΔH_f^o = 2.93 kcal/mol

HEXACYCLO[6.5.1.0^{2,7}.0^{3,11}.0^{4,9}.0^{10,14}]TETRADECANE
CAS RN   62415-11-6

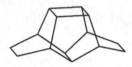

3,4,7-metheno-1H-cyclobuta[def]fluorene, dodecahydro

ΔH_f^o = 12.7 kcal/mol

HEXACYCLO[6.5.1.0^{2,11}.0^{3,14}.0^{4,10}.0^{9,13}]TETRADECANE
CAS RN   71871-53-9

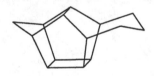

BISHOMOBIRDCAGE HYDROCARBON
1,5,2,4-ethanediylidenecyclopent[cd]azulene, dodecahydro

ΔH_f^o = 18.4 kcal/mol

HEXACYCLO[6.6.0.0^{2,11}.0^{3,7}.0^{4,9}.0^{10,14}]TETRADECANE
CAS RN   74999-12-5

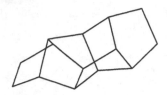

3,8,4,7-ethanediylidenecyclopent[a]indene, dodecahydro

ΔH_f^o = 7.9 kcal/mol

HEXACYCLO[7.3.1.1^{3,7}.0^{2,8}.0^{4,6}.0^{10,12}]TETRADECANE
CAS RN   61140-69-0

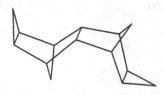

2,6:3,5-dimethanodicyclopropa[bg]naphthalene, dodecahydro
1aα, 2β, 2aβ, 3β, 3aα, 4aα, 5β, 5aβ, 6β, 6aα

ΔH_f^o = 63.9 kcal/mol

HEXACYCLO[7.4.1.0^{1,12}.0^{2,7}.0^{4,13}.0^{6,11}]TETRADECANE
CAS RN   50590-65-3

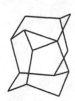

1,3-DEHYDRODIAMANTANE
2,1,4-ethanylylidene-1,6-methano-1H-cyclopropa[a]naphthalene, decahydro

ΔH_f^o = 28.9 kcal/mol

HEXACYCLO[8.3.1.0^{2,8}.0^{3,5}.0^{4,13}.0^{7,12}]TETRADECANE
CAS RN   50590-66-4

3,4-DEHYDRODIAMANTANE
3,5-DEHYDRODIAMANTANE
3,1,5-ethanylylidenecyclopropa[bc]acenaphthylene, dodecahydro

ΔH_f^o = 7.4 kcal/mol

HEXACYCLO[8.4.0.0^{2,12}.0^{3,7}.0^{4,9}.0^{8,11}]TETRADECANE
CAS RN   113719-48-5

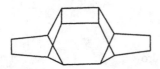

1,4,5,8-ethanediylidene-as-indacene, dodecahydro

ΔH_f^o = 20.6 kcal/mol

HEXACYCLO[9.2.1.0^{2,7}.0^{3,5}.0^{4,8}.0^{9,13}]TETRADECANE
CAS RN   74999-11-4

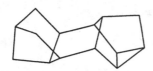

6,8-methano-1,2,4-metheno-s-indacene, dodecahydro
1α, 2α, 3aβ, 4α, 4aβ, 6α, 7aβ, 8α, 8aβ

ΔH_f^o = 22.5 kcal/mol

HEXACYCLO[9.2.1.0^{2,10}.0^{3,8}.0^{4,6}.0^{5,9}]TETRADECANE

CAS RN   96039-23-5   66289-73-4*

                 66289-74-5*   51966-13-3*

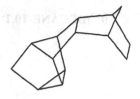

4,7-methano-2,3,8-methenocyclopent[a]indene, dodecahydro

ΔH_f^o = 34.6 kcal/mol

E. HEPTACYCLOTETRADECANES - $C_{14}H_{16}$

DISPIRO[CYCLOPROPANE-6,1'-PENTACYCLO[5.3.0.0^{2,5}.0^{3,9}.0^{4,8}]DECANE-10,1" - CYCLOPROPANE]
CAS RN   105883-34-9

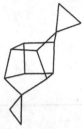

dispiro[cyclopropane-1,3'-[1,2,4]metheno[3H]cyclobuta[cd]pentalene-5'-(1'H),1"-cyclopropane], hexahydro

ΔH_f^o = 99.0 kcal/mol

HEPTACYCLO[6.6.0.0^{2,6}.0^{3,13}.0^{4,11}.0^{5,9}.0^{10,14}]TETRADECANE
CAS RN   17872-39-8

ISOGARUDANE
1,3,4,6-ethanediylidenedicyclopenta[cd,gh]pentalene, dodecahydro

ΔH_f^o = 10.6 kcal/mol

HEPTACYCLO[8.4.0.0^{1,11}.0^{2,7}.0^{3,5}.0^{4,8}.0^{9,14}]TETRADECANE
CAS RN   94359-42-9

1H-1,2,4:5,7a,8-dimetheno-s-indacene, decahydro

ΔH_f^o = 103.0 kcal/mol

HEPTACYCLO[8.4.0.0^{1,11}.0^{3,5}.0^{3,8}.0^{4,8}.0^{10,12}]TETRADECANE
CAS RN   73320-90-8

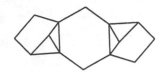

1H,4H,5H,8H-1,3a,8a:4a,5,7a-dimetheno-s-indacene, tetrahydro 1α, 3aα, 4aβ, 5β, 7aβ, 8aα

ΔH_f^o = 148.4 kcal/mol

HEPTACYCLO[8.4.0.0^{1,13}.0^{3,5}.0^{3,8}.0^{4,8}.0^{10,14}]TETRADECANE
CAS RN   73320-91-9

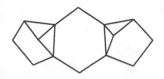

1H,4H,5H,8H-1,3a,8a:4a,7,7a-dimetheno-s-indacene, tetrahydro 1α, 3aα, 4aβ, 6β, 7αβ, 8aα

ΔH_f^o = 149.2 kcal/mol

HEPTACYCLO[8.4.0.0^{2,12}.0^{3,8}.0^{4,6}.0^{5,9}.0^{11,13}]TETRADECANE
CAS RN   75044-21-2
              13970-00-8*

BINOR-A
1,2,4:5,6,8-dimetheno-s-indacene, dodecahydro 1α, 2α, 3aβ, 4α, 4aα, 5β, 6β, 7aα, 8β, 8aβ

ΔH_f^o = 57.1 kcal/mol

HEPTACYCLO[8.4.0.0^{2,12}.0^{3,8}.0^{4,6}.0^{5,9}.0^{11,13}]TETRADECANE
CAS RN   13002-57-8
               13970-00-8*

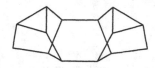

BINOR-S
1,2,4:5,6,8-dimetheno-s-indacene, dodecahydro 1α, 2α, 3aβ, 4α, 4aβ, 5α, 6α, 7β, 8α, 8aβ

ΔH_f^o = 60.3 kcal/mol

HEPTACYCLO[9.3.0.0^{2,5}.0^{3,13}.0^{4,8}.0^{6,10}.0^{9,12}]TETRADECANE
CAS RN   93569-21-2

GARUDANE
1,4-BISHOMO[6]PRISMANE
D$_{2h}$-BISHOMOHEXAPRISMANE
2,6,3,5-ethanediylidenecyclobut[jkl]as-indene, dodecahydro

ΔH_f^o = 54.4 kcal/mol

11
POLYCYCLOPENTADECANES
A. TRICYCLOPENTADECANES - C₁₅H₂₆

DISPIRO[4.1.5.3]PENTADECANE
CAS RN   13000-16-3

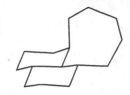

ΔH_f^o = - 52.4 kcal/mol

TRICYCLO[7.5.1.0³,⁸]PENTADECANE
CAS RN   92789-99-6

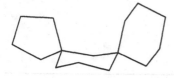

1,7-pentanonaphthalene, decahydro
1,10-ethano-1H-benzocyclononene, dodecahydro
ΔH_f^o = - 48.5 kcal/mol

TRICYCLO[9.3.1.0³,⁸]PENTADECANE
CAS RN   716-77-8

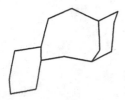

6,10-methanobenzcyclodecene, tetradecahydro
ΔH_f^o = - 42.4 kcal/mol

TRICYCLO[10.2.1.0^{2,11}]PENTADECANE
CAS RN   16539-04-1

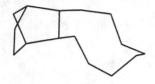

1,4-methanobenzocyclodecene, tetradecahydro

$\Delta H_f^o = -34.5$ kacl/mol

B. TETRACYCLOPENTADECANES - C₁₅H₂₄

DISPIRO[CYCLOPENTANE-1,2'-BICYCLO[2.2.1]HEPTANE-3',1"-CYCLOPENTANE]
CAS RN   41559-59-5

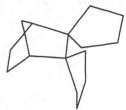

ΔH_f^o = - 22.1 kcal/mol

SPIRO[BICYCLO[6.1.0]NONANE-9,7'-BICYCLO[4.1.0]HEPTANE]
CAS RN   56424-53-9

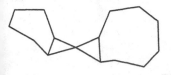

piro[bicyclo[4.1.0]heptane-7,9'-bicyclo[6.1.0]nonane]
ΔH_f^o = 12.4 kcal/mol

TETRACYCLO[6.4.3.0¹,⁸.0²,⁷]PENTADECANE

TETRACYCLO[6.4.3.0^{1,8}.0^{2,7}]PENTADECANE
CAS RN   63305-46-4

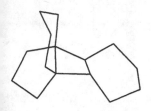

a,8b-propanobiphenylene, decahydro
H_f^o = -13.6 kcal/mol

TETRACYCLO[6.5.2.0^{4,15}.0^{11,14}]PENTADECANE
CAS RN   114925-13-2

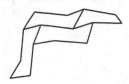

1H-cyclopenta[cd]pentalene, tetrahydro 2aα, 4aα, 7aα, 9aα, 9bα, 9cβ

ΔH_f^o = -42.9 kcal/mol

TETRACYCLO[6.5.2.0^{4,15}.0^{11,14}]PENTADECANE
CAS RN   114925-14-3

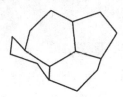

1H-cyclopenta[cd]pentalene, tetrahydro 2aα, 4aβ, 7aα, 9aβ, 9bα, 9cα

ΔH_f^o = -35.7 kcal/mol

TETRACYCLO[6.5.2.0^{4,15}.0^{11,14}]PENTADECANE
CAS RN   114857-88-4

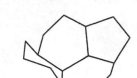

1H-cyclopenta[cd]pentalene, tetrahydro 2aα, 4aβ, 7aβ, 9aα, 9bβ, 9cα

ΔH_f^o = -31.4 kcal.mol

TETRACYCLO[6.6.1.0^{1,8}.0^{4,15}]PENTADECANE
CAS RN   80800-24-4

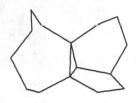

1H,4H-dicyclopenta[1,3:2,3]cyclopropa[1,2]cyclooctene, decahydro

ΔH_f^o = -16.2 kcal/mol

TETRACYCLO[6.6.1.0^{2,7}.0^{9,14}]PENTADECANE
CAS RN   107056-51-9

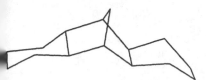

,10 -methanoanthracene, tetradecahydro

ΔH_f^o = -24.2 kcal/mol

TETRACYCLO[7.4.1.1^{1,11}.0^{5,14}]PENTADECANE
CAS RN   114857-87-3

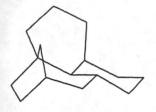

1H-6a,9-methanocyclohepta[de]naphthalene, dodecahydro 3aα, 6aβ, 9β, 10aα, 10bβ

H_f^o = -35.2 kcal/mol

237

TETRACYCLO[7.4.2.0^{1,10}.0^{3,8}]PENTADECANE
CAS RN   170931-91-6

3a,9-ethano-3aH-benz[f]indene, dodecahydro 3aα, 4aα, 8aα, 9α, 9aβ

ΔH_f^o = -37.6 kcal/mol

TETRACYCLO[7.4.2.0^{1,10}.0^{3,8}]PENTADECANE
CAS RN   171035-83-9

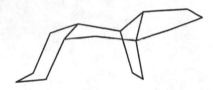

3a,9-ethano-3aH-benz[f]indene, dodecahydro 3aα, 4aα, 8aβ, 9α, 9aβ

ΔH_f^o = -30.4 kcal/mol

TETRACYCLO[7.4.2.0^{1,10}.0^{3,8}]PENTADECANE
CAS RN   171035-84-0

3a,9-ethano-3aH-benz[f]indene, dodecahydro 3aα, 4aβ, 8aα, 9α, 9aβ

ΔH_f^o = -31.5 kcal/mol

TETRACYCLO[7.6.0.0^{10,12}.0^{11,15}]PENTADECANE
CAS RN   53749-78-3

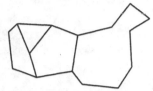

1H-cyclonona[a]cyclopropa[cd]pentalene, tetradecahydro

ΔH_f^o = 0.2 kcal/mol

TETRACYCLO[10.2.1.0^{1,9}.0^{3,8}]PENTADECANE
CAS RN   6902-95-0

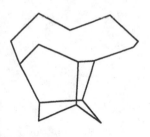

GIBBANE

1H-7,9a-methanobenz[a]azulene, dodecahydro 7α, 9α, 10aβ

ΔH_f^o = -37.4 kcal/mol

TETRACYCLO[10.2.1.0^{2,11}.0^{4,13}]PENTADECANE
CAS RN   100693-77-4

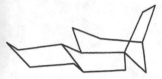

[5.1]TRIBLATTANE

5:2,4-dimethano-1H-cyclopentacyclodecene, dodecahydro

ΔH_f^o = 0.9 kcal/mol

TETRACYCLO[10.2.1.0^{2,11}.0^{3,10}]PENTADECANE
CAS RN   51532-36-6

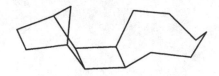

1,4-methanobenzo[3,4]cyclobuta[1,2]cyclooctene, tetradecahydro 1α, 4α, 4aβ, 4bα, 10aα, 10bβ

ΔH_f^o = -5.1 kcal/mol

TETRACYCLO[10.3.0.0^{2,6}.0^{7,11}]PENTADECANE
CAS RN   114925-11-0

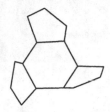

1H-trindene, tetradecahydro 3aα, 3bα, 6aα, 6bα, 9aα, 9bα

ΔH_f^o = -25.5 kcal/mol

TETRACYCLO[10.3.0.0^{2,6}.0^{7,11}]PENTADECANE
CAS RN   114925-12-1

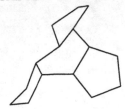

1H-trindene, tetradecahydro 3aα, 3bα, 6aα, 6bα, 9aα, 9bβ

ΔH_f^o = -31.6 kcal/mol

TETRACYCLO[10.3.0.0^{2,6}.0^{7,11}]PENTADECANE
CAS RN   114925-10-9

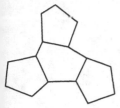

1H-trindene, tetradecahydro 3aα, 3bα, 6aα, 6bα, 9aβ, 9bβ

ΔH_f^o = -31.0 kcal/mol

TETRACYCLO[10.3.0.0^{2,6}.0^{7,11}]PENTADECANE
CAS RN   114925-07-4

H-trindene, tetradecahydro 3aα, 3bα, 6aα, 6bβ, 9aβ, 9bα

ΔH_f^o = -25.5 kcal/mol

TETRACYCLO[10.3.0.0^{2,6}.0^{7,11}]PENTADECANE
CAS RN   114925-09-6

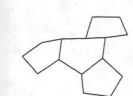

H-trindene, tetradecahydro 3aα, 3bα, 6aβ, 6bα, 9aα, 9bβ

H_f^o = -30.2 kcal/mol

TETRACYCLO[10.3.0.0^{2,6}.0^{7,11}]PENTADECANE
CAS RN   114925-06-3

1H-trindene, tetradecahydro 3aα, 3bβ, 6aα, 6bα, 9aα, 9bβ

ΔH_f^o = -33.6 kcal/mol

TETRACYCLO[10.3.0.0^{2,6}.0^{7,11}]PENTADECANE
CAS RN   114925-08-5

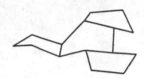

1H-trindene, tetradecahydro 3aα, 3bβ, 6aα, 6bβ, 9aβ, 9bα

ΔH_f^o = -34.5 kcal/mol

***cis,trans,cis*-TETRACYCLO[12.1.0.0^{4,6}.0^{9,11}]PENTADECANE**
CAS RN   1731-41-5
                      3105-37-1*

ΔH_f^o = 27.5 kcal/mol

---

trans, trans, trans-TETRACYCLO[12.1.0.0^{4,6}.0^{9,11}]PENTADECANE
CAS RN   23720-49-2
              3105-37-1*

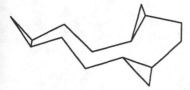

ΔH_f^o = 28.1 kcal/mol

TRISPIRO[4.0.4.0.4.0]PENTADECANE
CAS RN   29150-89-8

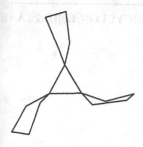

ΔH_f^o = -10.8 kcal/mol

C. PENTACYCLOPENTADECANES - C₁₅H₂₂

DISPIRO[BICYCLO[4.1.0]HEPTANE-7,1'-CYCLOPROPANE-2',7"-BICYCLO[4.1.0]HEPTANE]
CAS RN   71305-18-5
              71772-04-8*

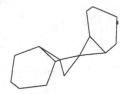

anti-1,2-di[7'-norcaranylidene]cyclopropane
ΔH_f^o = 51.0 kcal/mol

DISPIRO[BICYCLO[4.1.0]HEPTANE-7,1'-CYCLOPROPANE-2',7"-BICYCLO[4.1.0]HEPTANE]
CAS RN   71328-45-5
              71772-04-8*

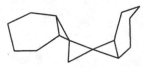

syn-1,2-di[7'-norcaranylidene]cyclopropane
ΔH_f^o = 69.2 kcal/mol

PENTACYCLO[6.5.1.1³,⁶.0²,⁷.0⁹,¹³]PENTADECANE
CAS RN   75172-86-0

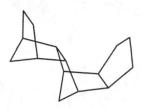

TETRAHYDROTRICYCLOPENTADIENE
4,9:5,8-dimethano-1H-benz[f]indene, dodecahydro 3aα, 4α, 4aβ, 5α, 8α, 8aβ, 9α, 9aα
ΔH_f^o = -3.3 kcal/mol

PENTACYCLO[6.5.1.1^{3,6}.0^{2,7}.0^{9,13}]PENTADECANE
CAS RN   66807-92-9

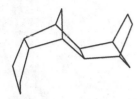

TETRAHYDROTRICYCLOPENTADIENE
4,9:5,8-dimethano-1H-benz[f]indene, dodecahydro 3aα, 4α, 4aβ, 5β, 8β, 8aβ, 9α. 9aα
ΔH_f^o = -1.9 kcal/mol

PENTACYCLO[6.5.1.1^{3,6}.0^{2,7}.0^{9,13}]PENTADECANE
CAS RN   51965-76-5

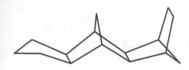

TETRAHYDROTRICYCLOPENTADIENE
4,9:5,8-dimethano-1H-benz[f]indene, dodecahydro 3aα, 4β, 4aα, 5α, 8α, 8aα, 9β, 9aα
ΔH_f^o = -5.1 kcal/mol

PENTACYCLO[6.5.1.1^{3,6}.0^{2,7}.0^{9,13}]PENTADECANE
CAS RN   75172-85-9

TETRAHYDROTRICYCLOPENTADIENE
4,9:5,8-dimethano-1H-benz[f]indene, dodecahydro 3aα, 4β, 4aα, 5β, 8β, 8aα. 9β, 9aα
ΔH_f^o = -6.3 kcal/mol

PENTACYCLO[6.5.1.1^{3,6}.0^{2,7}.0^{9,13}]PENTADECANE
CAS RN   75172-87-1

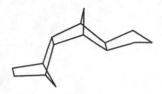

TETRAHYDROTRICYCLOPENTADIENE
4,9:5,8-dimethano-1H-benz[f]indene, dodecahydro 3aα, 4β, 4aβ, 5α, 8α, 8aβ, 9β, 9aα

ΔH_f^o = -5.3 kcal/mol

PENTACYCLO[7.4.1.1^{4,13}.0^{2,7}.0^{6,12}]PENTADECANE
CAS RN   41101-05-7

HOMODIAMANTANE
3,5,1,7-[1,2,3,4]butanetetrayl-1H-benzocycloheptene, decahydro

ΔH_f^o = -27.8 kcal/mol

PENTACYCLO[7.5.1.0^{2,8}.0^{3,7}.0^{10,14}]PENTADECANE
CAS RN   150375-98-2

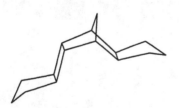

4,8-methanocyclopenta[3,4]cyclobuta[1,2-f]indene, tetradecahydro

ΔH_f^o = 4.2 kcal/mol

PENTACYCLO[7.6.0.0^{2,13}.0^{3,11}.0^{10,14}]PENTADECANE
CAS RN   88400-62-8 (-)

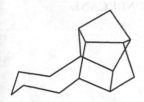

[5.1.1]TRIBLATTANE
1,3,5-methenocyclonona[cd]pentalene, tetradecahydro
ΔH_f^o = 34.7 kcal/mol

PENTACYCLO[9.2.1.1^{4,7}.0^{2,10}.0^{3,8}]PENTADECANE
CAS RN   39493-60-2

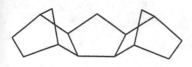

1,4:5,8-dimethano-1H-fluorene, dodecahydro
ΔH_f^o = - 6.1 kcal.mol

D. HEXACYCLOPENTADECANES - $C_{15}H_{20}$

DISPIRO[CYCLOPROPANE-5,1'-TETRACYCLO[6.2.1.0^{2,7}.0^{3,5}]UNDECANE-10',1"CYCLOPROPANE]
CAS RN   86301-93-1

dispiro[cyclopropane-1,6'(1'H)-[2,5]methanocycloprop[a]indene-7',1"-cyclopropane], octahydro
ΔH_f^o = 63.1 kcal/mol

HEXACYCLO[7.5.1.0^{3,13}.0^{5,12}.0^{7,11}.0^{10,14}]PENTADECANE
CAS RN   63127-44-6

3,4-methanocyclopenta[cd]pentaleno[2,1,6-gha]pentalene,tetradecahydro
[5]PERISTYLANE
C_{15}-HEXAQUINANE
ΔH_f^o = -5.6 kcal/mol

HEXACYCLO[9.2.2.0^{2,7}.0^{4,12}.0^{5,10}.0^{8,14}]PENTADECANE
CAS RN   86475-57-2

PENTASTERANE
1,7:2,6:3,5-trimethano-s-indacene, dodecahydro
ΔH_f^o = 0.3 kcal/mol

HEXACYCLO[9.3.1.0^{2,7}.0^{3,5}.0^{4,8}.0^{9,14}]PENTADECANE
CAS RN   189885-74-4

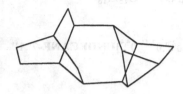

6,9-methano-1,2,4-methano-1H-benz[f]indene, dodecahydro

ΔH_f^o = 14.9 kcal/mol

PENTASPIRO[2.0.2.0.2.0.2.0.2.0]PENTADECANE
CAS RN   22748-12-5

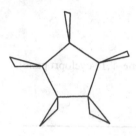

[5]ROTANE
[5.3]ROTANE
PENTACYCLOPROPYLIDENE

ΔH_f^o = 50.3 kcal/mol

E. HEPTACYCLOPENTADECANES - $C_{15}H_{18}$

DISPIRO[CYCLOPROPANE-1,4'-PENTACYCLO[6.3.0.0^{2,6}.0^{3,10}.0^{5,9}]UNDECANE-7',1"-CYCLOPROPANE]
CAS RN   156033-25-9

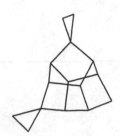

dispiro[cyclopropane-1,2'(4'H)-[1,3,5]methanocyclopenta[cd]pentalene-4',1"-cyclopropane], octahydro

ΔH_f^o = 63.3 kcal/mol

HEPTACYCLO[7.6.0.0^{2,7}.0^{3,14}.0^{5,12}.0^{6,10}.0^{11,15}]PENTADECANE
CAS RN   216673-78-8

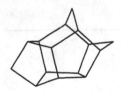

3,2,4-ethanylyidene-1,5methanodicyclopenta[cd,gh]pentalene, dodecahydro
ΔH_f^o = 5.2 kcal/mol

TRISPIRO[CYCLOPROPANE-1,3'-TETRACYCLO[3.3.1.0^{2,8}.0^{4,6}]NONANE-7,1"-CYCLOPROPANE-9,1'''-CYCLOPROPANE]
CAS RN   125263-26-5

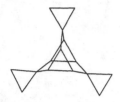

TRISPIROCYCLOPROPANETRIASTERANE
trispiro[tetracyclo[3.3.1.0^{2,8}.0^{4,6}]nonane-3,1':7,1":9,1''' triscyclopropane]cyclopropane
ΔH_f^o = 48.8 kcal/mol

TRIVIAL NAME INDEX